Communicating Endangered Species

Extinction, News and Public Policy

Edited by Eric Freedman, Sara Shipley Hiles and David B. Sachsman

Routledge
Taylor & Francis Group
LONDON AND NEW YORK

earthscan
from Routledge

First published 2022
by Routledge
2 Park Square, Milton Park, Abingdon, Oxon OX14 4RN

and by Routledge
605 Third Avenue, New York, NY 10158

Routledge is an imprint of the Taylor & Francis Group, an informa business

British Library Cataloguing-in-Publication Data
A catalogue record for this book is available from the British Library

Library of Congress Cataloging-in-Publication Data
Names: Freedman, Eric, editor.
Title: Communicating endangered species : extinction, news and public policy / edited by Eric Freedman, Sara Shipley Hiles, and David B. Sachsman.
Description: Milton Park, Abingdon, Oxon ; New York, NY : Routledge, 2022. | Series: Routledge studies in environmental communication and media | Includes bibliographical references and index.
Identifiers: LCCN 2021007495 (print) | LCCN 2021007496 (ebook)
Subjects: LCSH: Endangered species. | Communication in the environmental sciences.
Classification: LCC QH77.3.C65 C66 2022 (print) | LCC QH77.3.C65 (ebook) | DDC 333.95/22—dc23
LC record available at https://lccn.loc.gov/2021007495
LC ebook record available at https://lccn.loc.gov/2021007496

ISBN: 978-0-367-48621-1 (hbk)
ISBN: 978-1-032-04544-3 (pbk)
ISBN: 978-1-003-04195-5 (ebk)

Typeset in Bembo
by Apex CoVantage, LLC

To environmental journalists around the world whose efforts continue to bring the extinction and biodiversity crisis to the attention of the public and policy makers.

Contents

Figures

Tables

Editors and contributors

Editors

Eric Freedman is Knight Chair, Director of the Knight Center for Environmental Journalism, and a professor in the School of Journalism at Michigan State University. He also directs MSU's Capital News Service, a public affairs reporting practicum on state government, policy, and politics. He has led study abroad programs on Australia Media, Environment, Culture & Tourism, Reporting in the British Isles, Mass Media in the United Kingdom, and Media and Environment in Scotland. As a Fulbright Scholar, he taught journalism and conducted research in Uzbekistan, Lithuania, and the Republic of Georgia. As a journalist, his interests include public lands, habitat and diversity, invasive species, ecotourism, forests, international transborder environmental problems, and environmental litigation. His books include *Biodiversity, Conservation and Environmental Management in the Great Lakes Basin*; *Critical Perspectives on Journalistic Beliefs and Actions: Global Experiences*; *Environmental Crises in Central Asia: From Steppes to Seas, from Deserts to Glaciers*; *After the Czars and Commissars: Journalism in Authoritarian Post-Soviet Central Asia*; *African Americans in Congress: A Documentary History*; and *Presidents and Black Americans: A Documentary History*. During his 20-year newspaper-reporting career, Freedman covered public affairs and legal affairs in New York and Michigan, winning a Pulitzer Prize for coverage of a legislative corruption scandal.

Sara Shipley Hiles is an associate professional practice professor at the University of Missouri School of Journalism. Her areas of interest include environmental, science, and health reporting; using multimedia and social media; and investigative reporting. Hiles started her journalism career at the *Times-Picayune* in New Orleans while still a student at Loyola University. Later, she moved to the *Statesman-Journal* in Salem, Oregon, where she realized her dream of covering the environmental beat, including traveling to Iceland to cover "Keiko" the killer whale's journey to freedom. At the *Courier-Journal* in Louisville, Kentucky, she co-wrote investigative projects on mentally ill jail inmates and on railroad workers exposed to toxic solvents on the job. At the *St. Louis Post-Dispatch*, she reported on the endangered Ozark hellbender salamander, floodplain development, and the scourge of "popcorn lung" disease suffered by factory workers across the country. Her freelance work has been published in the *Christian Science Monitor*, the *Boston Globe*, *Mother Jones*, National Public Radio, and elsewhere. She is co-author of a book on Hurricane Katrina and New Orleans. Hiles has taught at Western Kentucky University and Bowling Green State University in Ohio and won awards and recognition for teaching, feature writing, digital journalism, and investigative reporting.

David B. Sachsman holds the George R. West, Jr. Chair of Excellence in Communication and Public Affairs at the University of Tennessee at Chattanooga. Previously he served as dean of the School of Communications at California State University, Fullerton, and as chair of the Department of Journalism and Mass Media at Rutgers University. His research and writing on environmental journalism began in the 1970s with his Stanford University dissertation on the influence of public relations on environmental coverage. Sachsman's works include *Environmental Risk and the Press* and *The Reporter's Environmental Handbook*, and continue to this day with *Environment Reporters in the 21st Century*, a book chapter, "Environmental Reporters," and the *Routledge Handbook of Environmental Journalism*, which were written or edited with JoAnn Myer Valenti. In 2005, he led the team charged with evaluating the US Agency for International Development's environmental communication and education projects in more than 30 countries. He serves as the managing editor of the international journal *Applied Environmental Education and Communication*. He is also a journalism historian and the director of the annual Symposium on the 19th Century Press, the Civil War, and Free Expression, which he founded in 1993. He is an editor of eight books of readings on the 19th century press, drawn from papers originally presented at the symposium.

Contributors

Dorothea Born is a lecturer at the Department for Science and Technology Studies at the University of Vienna where she earned a PhD for investigating how climate change was visually made a concern in two popular science magazines, *National Geographic* and *GEO*. She earned a master's degree in ecology, researching the impacts of climate change on Arctic permafrost soils. Her research centers on visual representations of climate change.

John Carr is an urban and legal geographer whose work focuses on the intersections of urban geography, law, planning, and human and non-human environments. His research seeks to address questions that can be answered only by bridging the gaps among different disciplines. Projects include work addressing how environmental laws both separate and complicate the boundaries between human-built environments and the more-than-human. He is a senior lecturer with the Environment and Society Group at the University of New South Wales.

Harriet Croome is a student in the University of Birmingham's Forest Edge Doctoral Scholarship Programme, studying the effects of conservation initiatives on the relationship between human and non-human forest users in Kenya. She earned her MSc in science communication at the University of the West of England, Bristol. She has worked in public engagement and program coordination for wildlife conservation NGOs in the United Kingdom and Madagascar and is a qualified field guide with the Field Guides Association of Southern Africa.

Kerrie Foxwell-Norton is an associate professor in the School of Humanities, Languages, and Social Science at Griffith University, Australia, where she is the communication and journalism disciplinary leader. She is a member of the Griffith Centre for Social and Cultural Research and leads a research team within the Griffith Climate Action Beacon. Her research focuses on environmental communication, media, and journalism with a particular interest in the communication of coastal and marine environments. Her books include *The Local and the Digital in Environmental Communication*; *Environmental Communication and Critical Coastal Policy: Communities, Culture and Nature*; and *Journalism and Climate Crisis: Public Engagement, Media Alternatives*.

Caleb R. Hickman is a supervisory biologist for the Eastern Band of Cherokee Indians. Most of his work follows an adaptive social-ecological process where, through research, restoration, and planning, he works to improve conditions for the ecosystem and the Cherokee people alike. This process involves influencing decision-makers through significant public and government engagement. He has authored and co-authored numerous publications at the edge of ecology and social science. Prior to working for the tribe, he earned his PhD in zoology from the University of Wisconsin-Madison, master's degree in biology from Missouri State University, and bachelor's degree in biology from Missouri Valley College.

Elaine Jeffreys is a professor in International Studies, Faculty of Arts and Social Sciences, University of Technology Sydney. She earned her PhD in political science at the University of Melbourne. Recent books include *Governing HIV in China*, *New Mentalities of Government in China*, *Sex in China*, and *Celebrity Philanthropy*.

Leandro Jerusalinsky is an environmental analyst at the Chico Mendes Institute for Biodiversity Conservation, linked to the Brazilian Ministry of Environment, where he directs the National Center for Research and Conservation of Brazilian Primates. He earned a bachelor's degree in biological sciences, a master's in genetics and molecular biology, and a PhD in biological sciences – zoology in Brazil. He is the president of the Latin-American Society of Primatology, vice-chair of the IUCN SSC Primate Specialist Group for Brazil and the Guianas, and a member of the IUCN SSC Conservation Planning Specialist Group. His focus is on neotropical primate conservation.

Daphné Kerhoas is the conservation science and higher education manager at the Bristol Zoological Society where she leads the Society's Philippines conservation program. She earned her PhD in primatology from the Max Planck Institute of Evolutionary Anthropology and the University of Leipzig. She is an IUCN SSC Primate Specialist Group member for the South-East Asia section and a reviewer for the PSGB Conservation Working Party. Her focus is on assessing conservation initiatives.

Joe Kerkvliet is professor emeritus of Economics at Oregon State University and senior economist at the Conservation Economics Institute. His most recent research has focused on conservation on private land, the workings and performance of the Endangered Species Act, and economic impacts of unconventional oil and gas development. He lives in Oregon, where he hunts mushrooms and steelhead, and in Montana, where he hunts deer, elk, and trout. He designs and contributes to land and water restoration projects in both states.

Alexander Killion is a conservation scientist and a PhD candidate at the University of Michigan School for Environment and Sustainability. He also serves as a National Aeronautics and Space Administration Research Fellow. A former National Socio-Environmental Synthesis Center Graduate Research Fellow, his research is transdisciplinary and focuses on human–environment interactions, biodiversity conservation, and sustainable development. He earned his master's degree in fisheries and wildlife at Michigan State University.

William Konstant is an advisor to the Margot Marsh Biodiversity Foundation and the Mohamed bin Zayed Species Conservation Fund and a member of the IUCN SSC Primate Specialist Group. Since the early 1980s, he has helped develop and support projects in the New World tropics, Africa, Madagascar, and Asia, worked with international nongovernmental organizations, and developed conservation programs at North American zoos. He is a former co-chair of the Association of Zoos and Aquariums' Field Conservation Committee. His priorities include supporting development of an action plan for Africa's endangered red colobus monkeys and supporting the newly created African Primatological Society.

Bill Kovarik is an historian whose work is located at the intersection of technology, communications, and the environment. His books include *Revolutions in Communication, Mass Media and Environmental Conflict* (with Mark Neuzil), and the *Ethyl War*. He earned a PhD at the University of Maryland.

His professional background includes wire services, daily newspapers, and science and environmental publications. He teaches media history, environmental history, and communications law at Radford University.

Gavin Lamb is a postdoctoral fellow in the Center for Multilingualism in Society Across the Lifespan at the University of Oslo. His research is in the areas of interactional sociolinguistics and ecolinguistics, and explores how language, culture, and communication shape diverse human relationships with threatened species and places. He has published and co-authored several articles investigating the role of intercultural communication in sea turtle conservation and tourism settings in Hawai'i. He received his PhD in Second Language Studies at the University of Hawai'i at Mānoa.

Christian Langpap is a professor in the Department of Applied Economics at Oregon State University. His research focuses on the economics of species conservation under the Endangered Species Act and on the role of environmental nonprofits in enforcement of environmental regulation and provision of environmental public goods. He earned his PhD in agricultural and resource economics at Oregon State University.

Libby Lester is the director of the Institute for Social Change and a professor of Media and Communication at the University of Tasmania, Australia. Her research focuses on how issues are raised, understood, and responded to in public debate; and she is recognized internationally for her work on environmental communications. Before joining the university, she worked as a journalist, reporting on social, political, and environmental issues for major Australian newspapers and magazines. Her latest book is *Global Trade and Mediatised Environmental Protest: The View from Here.*

Gráinne McCabe is head of field conservation and science for the Bristol Zoological Society, where she directs global conservation programs in 10 countries and oversees the society's higher education programs, including six university degree programs. She leads two conservation programs directly: Western Lowland Gorilla Conservation Project in Equatorial Guinea and Sanje Mangabey Conservation Project in Tanzania. She is a member of the IUCN SSC Primate Specialist Group, Africa Section, and the IUCN BIOPAMA Regional Advisory Committee for West and Central Africa, and is a trustee for the Cameroonian primate sanctuary, Ape Action Africa.

Tema Milstein is an associate professor of Environment & Society at the University of New South Wales in Sydney. A former journalist who became an academic, her work looks at ways culture, society, and communication inform – and are informed by – environmental relations. Her research focuses on ways cultural meaning systems shape our ecological understandings, identities, and actions. Her recent books are *Routledge Handbook of Ecocultural Identity* and *Environmental Communication Pedagogy and Practice.*

Russell A. Mittermeier is the chief conservation officer of Global Wildlife Conservation and chair of the IUCN SSC Primate Specialist. Prior to that, he served as president of Conservation International for 25 years. He has been working on biodiversity conservation in tropical forests for the past 50 years, with an emphasis on primate conservation. He originated the concept of the Top 25 in 1999 in collaboration with colleagues and has been promoting it ever since. He also created the concept of primate-watching and primate life-listing, based on a bird-watching model. His publications include 41 books and more than 700 scientific and popular publications.

Mark Neuzil is a professor and chair of the Department of Emerging Media at the University of St. Thomas-Minnesota. He is the author, co-author, or editor of eight books, mostly on environmental issues, and is a frequent writer and speaker on the environment. His most recent book is *Canoes: A Natural History in North America*. He has reported for the Associated Press, the *Minneapolis Star Tribune*, and several other newspapers. He was a Fulbright Scholar in the Czech Republic and a Distinguished Visiting Professor in Liberal Arts at the University of Minnesota-Morris.

Michael A. Patten is a professor of Terrestrial Ecology and Nature Management at Nord University in Steinkjer, Norway. Prior to that, he was a Presidential Professor at the Oklahoma Biological Survey at the University of Oklahoma, in Norman, where he specialized in conservation biology and applied ecology.

Jonah Ratsimbazafy received his PhD at the State University of New York at Stonybrook in 2002 and did his higher degree research at the University of Antananarivo, Madagascar, in 2011. He is a faculty lecturer at the University of Antananarivo; co-vice chair of the Madagascar section of the IUCN/SSC Primate Specialist Group; director of the Houston Zoo, Madagascar Program; president of the Madagascar Primate Research Group; and a member of the Malagasy Academia, the African Academy of Sciences, and the World Academic of Sciences. He is president of the International Primatological Society.

Kim E. Reuter is a senior advisor on climate change in the UK Foreign, Commonwealth, and Development Office. A biologist by training and expert in international development, she has consulted for the World Bank, WWF, and Global Wildlife Conservation, served as the senior technical director for the Gaborone Declaration, and led two charities as the head executive. She has worked on conservation projects in 16 African countries and authored more than 30 scientific publications. She is a 2016 National Geographic Explorer and 2020 Churchill Fellow, and volunteers as a board member for three nonprofit organizations.

Anthony B. Rylands is the primate conservation director at Global Wildlife Conservation. He began his career in 1976 at the National Institute for

Amazon Research in Manaus, Brazil, and earned his doctorate at the University of Cambridge for a study of the behavior and ecology of marmosets and lion tamarins in 1982. From 1986 to 2003, he was a professor of vertebrate zoology at the Federal University of Minas Gerais, Brazil, and, from 2000 to 2017, a researcher at Conservation International. He is a member of the Brazilian Academy of Sciences and deputy chair of the IUCN SSC Primate Specialist Group.

Christoph Schwitzer is the director of the Dublin Zoo and former chief zoological officer and deputy chief executive of the Bristol Zoological Society. He spent two years in Madagascar leading a lemur research and conservation program. He earned his PhD in zoology from the University of Cologne and has been a visiting professor at the University of the West of England since 2013. He is the chair of council of the British and Irish Association of Zoos and Aquariums, the deputy chair of the IUCN SSC Primate Specialist Group, and the vice president of the Lemur Conservation Council.

Jason Shogren is the Stroock Chair of Natural Resource Conservation and Management in the Department of Economics at the University of Wyoming, his alma mater. He studies the behavioral and institutional underpinnings of environmental and economic policy. He is a foreign member of the Royal Swedish Academy of Sciences, and served as professor to King Carl XVI Gustaf of Sweden. He worked with the Intergovernmental Panel on Climate Change, the Council of Economic Advisers in the White House, and Wyoming's Environmental Quality Council. He is a Fellow of the Association of Environmental & Resource Economists, the Agricultural & Applied Economics Association, and the Beijer Institute of Ecological Economics.

Brenda D. Smith is a conservation biologist with the Oklahoma Natural Heritage Inventory, a program of the Oklahoma Biological Survey at the University of Oklahoma. Her research employs general natural history, ecology, and conservation biogeography to inform extinction risk assessments of vertebrates and invertebrates. Her publications include *Is Extinction Forever?* (*Public Understanding of Science*) and the book *Dragonflies at a Biogeographical Crossroads: The Odonata of Oklahoma and Complexities Beyond Its Borders*.

Michael Scott Smith is a PhD researcher in illegal wildlife trade news framing in the School of Humanities, Languages and Social Science at Griffith University, Australia. A freelance environmental reporter, he has reported for *National Geographic*, Mongabay, and ABC Open on the illegal trade in rhino horn in Vietnam, rhino poaching in South Africa, bleaching events on the Great Barrier Reef, and wildlife conservation issues in Vietnam and Australia. He earned his Master of Journalism at the University of Queensland and did a Master of Arts Research thesis on Vietnamese environmental nongovernmental organizations' rhino horn demand-reduction media campaigns at Griffith.

Karen B. Strier is Vilas Research Professor and Irven DeVore Professor of Anthropology at the University of Wisconsin-Madison. She coordinates the longest-ongoing field study of the critically endangered northern muriqui in the Atlantic Forest of southeastern Brazil, where she has trained more than 75 Brazilian students. She has received many national and international awards and recognition for her scientific contributions in primate behavioral ecology and for her impact on primate conservation. She is the author of *Faces in the Forest: The Endangered Muriqui Monkeys of Brazil* and *Primate Behavioral Ecology* (6th edition). She is a former president of the International Primatological Society.

Shem D. Unger is an associate professor of Biology at Wingate University, where he teaches organismal biology courses. The majority of his research is on the conservation of reptiles and amphibians. His primary research interests include studying the demography and behavior of the fully aquatic eastern hellbender salamander, conservation genetics, and also using non-invasive sampling tools for conducting wildlife surveys.

Ryan Wallace is a PhD candidate in the School of Journalism & Media at the University of Texas at Austin and earned his master's degree at California State University, San Marcos. With an interdisciplinary background spanning the life, physical, and social sciences, his research interests center around mediated science communications with a focus on key issues such as the Anthropocene, scientific misinformation, new media, and development in Latin America. His research has been published in *#ISOJ: The Journal of the International Symposium on Online Journalism*, as well as other texts.

Amanda D. Webber is an anthropologist at the Bristol Zoological Society where she co-leads the Northern Madagascar conservation project alongside a lecturing position. Previously, she was a lecturer at Oxford Brookes University and the University of Bristol. Her work focuses on human–wildlife interactions, and she is a member of the IUCN Primate Specialist Group Section for Human Primate Interactions; PSG Section for Madagascar; and the Stork, Ibis, and Spoonbill Specialist Group.

Clare Wilkinson is an associate professor in Science Communication and co-director of the Science Communication Unit at the University of the West of England, Bristol. Her main research focus is public engagement, including how different disciplines, researchers, and public participants are mobilized to engage. She earned her PhD in sociology at the University of Plymouth and has published in journals including *Science Communication*, *Journalism Practice*, and *Public Understanding of Science*. She co-authored the book *Creative Research Communication: Theory and Practice* and is the series editor for Contemporary Issues in Science Communication published by Bristol University Press.

Elizabeth A. Williamson is an honorary senior research fellow at the University of Stirling. Between 1982 and 2004, she studied or surveyed gorillas and chimpanzees in Cameroon, Central African Republic, Democratic Republic of Congo, Gabon, and Liberia, including six years as director of the Karisoke Research Centre in Rwanda. She serves as a Red List authority coordinator for the IUCN SSC Primate Specialist Group and on the executive committee of the PSG's Section on Great Apes.

Jessica Wise is a conservation biologist and science communicator. After completing a master's degree in applied animal behavior and animal welfare at the University of Edinburgh, she became the conservation engagement manager at the Royal Zoological Society of Scotland, where she engaged and connected with a variety of audiences. As the founder of ConservationWise, she works with international projects providing fieldwork and placement advice to students and early career conservationists and writes an eco-travel blog to inspire, inform, and instigate societal change by a non-technical audience.

Acknowledgments

We appreciate the research assistance of Environmental Science graduate students Mary Feely, Rachel Head, McKenzie Whitten, and Emily Davis at the University of Tennessee at Chattanooga and the proofreading assistance of undergraduates Logan Bry and Alexandra Swanson of the Professional and Public Writing Program at Michigan State University. We also appreciate the work of Professor Mark Neuzil, Chair of the Department of Journalism and Mass Communication at the University of St. Thomas, in helping to conceive this project.

1 Exploring the terrain

Connecting communication, public policy, and the brink of extinction

Eric Freedman, David B. Sachsman and Sara Shipley Hiles

It is bone-chilling to know the precise date a species disappeared. It is even more ghastly to look upon the place where it happened and know that nobody knew or cared at the time what had transpired and why. There is little to memorialize the deliberate environmental disaster that occurred at the now-abandoned site of the Beaumaris Zoo in Hobart, the capital of Tasmania. The disaster that occurred there was not unprecedented, but it was a harbinger of a global future with less and less biodiverse animal and plant life.

For 14 years, the zoo had "introduced a wide range of exotic and native fauna to the curious and inquisitive people of Hobart," including the thylacine, or Tasmanian tiger. Australian newspapers, politicians, and farmers had long vilified the thylacine (*Thylacinus cynocephalus*), falsely blaming it as a predator that could wipe out flocks of sheep and encouraging bounty hunting of the continent's largest carnivorous marsupial. As the online Thylacine Museum explains, "Sadly, out of ignorance, irrational fear, and largely just because it was perceived as an economic threat, a concerted war of extermination was waged against the species" (Campbell, n.d.).

The Beaumaris Zoo had been built at the location of a 19th-century sandstone quarry worked by prisoners transported as punishment from England to Australia. After the zoo closed in 1937, the site became rough, weed-covered ground, and its hourglass-shaped concrete duck pond was drained of water and layered with debris. Beyond a gray toilet block stood the cage where the world's last known thylacine had died. Its name: Benjamin. The date: September 7, 1936, only months after the Tasmanian government bestowed much-belated legal protection on the species. It was too little by far, too late by far for Benjamin, which became what scientists call an endling, the last survivor of its species. Today, Benjamin merits only a few lines of type, an etching, and a childlike metal caricature on a small metal plaque at the zoo site for passersby to read. All that is left of this doomed endling itself are a few black-and-white photos and a brief film clip showing it pacing in its cage.

Endling. A word with finality.

Martha, the world's last known passenger pigeon (*Ectopistes migratorius*) was an endling whose final resting place is in Washington, DC. She had been a celebrity at the Cincinnati Zoo until September 1, 1914, when her death eliminated the last vestige of a seemingly boundless species so numerous that huge flocks sometimes darkened the skies. Along the Lake Michigan coast near Petoskey, Michigan, for instance, as many as 50,000 a day were reportedly shot in 1878, according to contemporary press reports. Writing about their declining numbers, Fuller (2014) said the passenger pigeon population had been dropping for years before Martha's death as the public largely remained indifferent because millions of birds remained (see Figure 1.1).

After Martha's death, her body was encased in a block of ice for shipping to the Smithsonian Museum of Natural History for scientific study. She was mounted and placed on a small branch, now fastened to a block of Styrofoam, and paired with a male passenger pigeon that had died in 1873 in Minnesota. The two birds had no connection with each other during life and were mated only for public display. They now spend virtually all their time in a nondescript storage locker next to one containing birds that US President Theodore Roosevelt had shot and studied as a boy.

On the opposite side of the globe is evidence of another extinction housed in a small, dingy museum in the bleak Central Asian city of Nukus, in the far west

Figure 1.1 Martha, the last known passenger pigeon. She died in the Cincinnati (Ohio) Zoo on September 1, 1914.

Source: From the Smithsonian Institution.

of the former Soviet republic of Uzbekistan. This endling is mounted in a drab display case: a Caspian tiger (*Panthera tigris virgata*) felled by a hunter's bullet in the 1950s, according to a plaque. Unlike Martha and Benjamin, the stuffed tiger had no name before its death. Many scientists disagree with the museum's claim that this particular Caspian tiger was, indeed, the final survivor, another endling. Some assert that the last one had died in northern Iran, or northeastern Afghanistan, or perhaps elsewhere. But there's no doubt the species is extinct today.

To some people, such physical remnants of doomed species seem macabre, reflecting a weird fascination with avoidable-turned-inevitable large-scale deaths. But, like a cemetery visit to read ancient headstones, endlings can teach us lessons, such as how easily abundance can disappear into nothing and how quickly carelessness or callousness can wipe a species off the face of the Earth.

It is also important to emphasize that none of those three extinctions drew significant media attention, if any, when they occurred. Rather they vanished with little or no public notice, as did the dodo of Mauritius – last scientifically reported seen in the 1600s – and nine species of the New Zealand moa – last scientifically reported in the 1800s. As did thousands of other extinct species. As will thousands more.

Yes, extinction certainly isn't new. However, the pace is accelerating globally. The International Union for Conservation of Nature (IUCN) classifies more than 30,000 species as threatened. The causes are many, as this book makes clear, including climate change, overdevelopment, human exploitation, disease, overhunting, habitat destruction, and predators. The willingness and the ability of ordinary people, governments, scientists, nongovernmental organizations, and business to slow this deeply disturbing acceleration are uncertain. Meanwhile, researchers around the world are laboring to better understand and communicate the possibility and implications of extinctions and to discover effective tools and public policies to combat the threats to species survival.

This multidisciplinary environmental communication book takes a distinctive approach by connecting how media and culture depict and explain endangered species with how policy makers and natural resource managers can or do respond to these challenges in practical terms. It draws on the latest research by authors who are a mix of social scientists, communication experts, and natural scientists from four continents. Each chapter includes mass media and policy aspects.

Threats loom large

Biodiversity and species survival confront a wide array of perils, from broad climate change to habitat destruction, to invasive intruders, to diseases. Hoping to improve the success of conservation projects such as land, water, and species management, researchers in the United States, United Kingdom, and Switzerland classified threats into 11 categories (Salafsky et al., 2008):

- residential and commercial development;
- agriculture and aquaculture;

- energy production and mining;
- transportation and service corridors;
- natural resource use, such as logging, hunting, and fishing;
- human intrusions and disturbance, including recreational activities;
- natural system modifications, such as dams and fire suppression;
- invasive and other problematic species;
- pollution;
- geological events, such as volcanoes, earthquakes, and landslides;
- climate change and severe weather.

Some threats, of course, are beyond human control, particularly geological events. However, most threats, including discharge of pollutants, logging and mining practices, and construction of dams and roads, can be mitigated.

Today, the mass media generally present two basic types of bad-news, parade-of-horribles extinction stories.

One set of stories from journalists, environmental communicators, and advocacy groups takes a global view as they cite the latest warnings from scientists that "civilization" could eliminate species after species. These startling, frightening numbers are mind-boggling in scope and, thus, deemed newsworthy. Here's an example: A report from the Intergovernmental Science-Policy Platform on Biodiversity and Ecosystem Services, which works with the United Nations Environment Programme, cautions, "Nature is declining globally at rates unprecedented in human history – and the rate of species extinctions is accelerating, with grave impacts on people around the world now likely." The report projects that about 1 million plant and animal species "are now threatened with extinction, many within decades, more than ever before in human history" (2019). The press picked up on the report, seizing on the sensational as reflected by a *Washington Post* headline using the phrase "face of an extinction crisis." The article goes on to say:

> The North Atlantic right whale is "one step from extinction," an international group of governments and scientists declared Thursday, slightly more than a month after President [Donald] Trump lifted restrictions on commercial fishing in a key area of the whale's habitat . . . More than 1 in 4 species around the world are threatened with extinction. Lemurs in Madagascar have plummeted from hunting. European hamsters have declined because of development. And "the world's most expensive fungus," the caterpillar fungus on the Tibetan Plateau, has been nearly wiped out by overharvesting for use in traditional Chinese medicine.
>
> (Fears, 2020)

The International Union for Conservation of Nature, which maintains the authoritative and depressingly long Red List of Threatened Species, also presents such grim – and newsworthy – forecasts. On a macro level, IUCN (2020) says more than 30,000 of the 112,400-plus plant and animal species it has assessed are threatened. That dire prediction includes about a quarter of land mammal species, a third of marine mammal species, and a third of amphibian species that the organization says face extinction in the foreseeable future, meaning the next 10–100 years. Another study places 515 of 29,400 terrestrial vertebrate species with fewer than 1,000 remaining individuals on the edge of extinction (Ceballos, Ehrlich, & Raven, 2020). Those species include the Sumatran rhino, Clarion island wren, Española giant tortoise, and harlequin frog.

An IUCN study (2019a) titled "A Miniature World in Decline: European Red List of Mosses, Liverworts, and Hornworts" estimated that 22.5% of 1,817 bryophyte species native to Europe face extinction, largely from land modifications, climate change, fires, and conversion of forests to plantation woodland. That assessment said 2% of European medicinal plants, 8% of aquatic plants, 16% of crop-wild relatives, 20% of ferns and lycopods, and 42% of trees are threatened as well. A companion report on native European tree species found that 42% of the 454 species assessed are critically endangered, endangered, or vulnerable due to invasive species, disease, deforestation, wood harvesting, urban development, and livestock farming, among other dangers (IUCN, 2019b). Few readers would recognize many of the species named in those reports.

In addition to global stories, many stories focus primarily on charismatic or familiar species, ones that resonate with the public at large or ones that can be individualized – even, ideally, given human names like Martha and Benjamin. In 2010, scientists bestowed the name Flex on a 13-year-old western gray whale tagged off Russia's Pacific coast. Flex was among only about 136 remaining members of that critically endangered species. A northern bald ibis, the Middle East's most critically endangered bird species, was named Julia when she was tagged after being shot while migrating across northern Saudi Arabia.

The *New York Times* ran an account of the death in a Hawaiian terrarium of a 14-year-old land snail named George, believed to be the last of its species. "Scientists estimate that dozens of species go extinct each day, but few receive this kind of news media attention on their way out," the article accurately noted of this endling. The article quoted the founder of a local snail protection program, who said, "Naming George probably boosted his standing . . . You anthropomorphize it and people pay attention" (Jacobs, 2019). Similarly, international media took notice when the last known female Yangtze giant softshell turtle died in China after a failed artificial insemination attempt. Only three others were confirmed to exist at the time: a male in a zoo and two of unknown gender in Vietnamese lakes (Mongabay.com, 2019). In Brazil, two songbirds went extinct, and in Kenya, the last known male northern white rhino died at a wildlife sanctuary, leaving only two females alive (Rice, 2019).

Is it foolish to name individual members of near-gone species? While sentimentality won't save them, naming them does serve a purpose: It is a way to communicate to the public and policy makers about the extinction crisis and to help them identify with the situation.

The future sounds depressingly sad, according to many media accounts. Looking gloomily ahead, a news report about one study warns that climate change is responsible for the disappearance of the familiar bumble bee at rates "consistent with a mass extinction" (Rice, 2020). A major US newspaper quotes the Center for Biological Diversity saying dramatically, "We're currently experiencing the worst spate of die-offs since the loss of the dinosaurs 65 million years ago" (Rice, 2019). Most polar bears "will probably be gone by 2100," with just a few surviving in the Arctic, a newspaper account of another study says (Dickie, 2020).

Not surprisingly, serious disagreements exist over the quality and implications of news coverage of endangered species, even by reputable and respected press organizations. The blunt title of one study by Canadian and US researchers reflects such critiques: "Inaccurate and Biased Global Media Coverage Underlies Public Misunderstanding of Shark Conservation Threats and Solutions." That analysis of a decade's worth of reports in the "popular press" found that the controversial practice of finning was mentioned in two-thirds of news articles about threats to sharks; meanwhile, a greater threat – overfishing – was mentioned in fewer than half of articles. The study comments, "We find that shark conservation threats and solutions are frequently presented in an over-simplified, biased, or factually inaccurate manner that would likely contribute to widespread public misunderstanding about these topics" (Shiffman et al., 2020, p. 5).

Similarly, Yong (2019) sought to counter the "drumbeats of doom" about an "insect apocalypse" or "Armageddon" in which all insects could disappear within a century. He writes, "The data on insect declines are too patchy, unrepresentative, and piecemeal to justify some of the more hyperbolic alarms," including stories published in the *New York Times* and the *Guardian*. He adds, "At the same time, what little information we have tends to point in the same worrying direction." Next, he asks this key question: "How, then, should we act on that imperfect knowledge?"

In reality, most threatened and endangered species slide under the media radar. That does not mean their plights are not dire. Nor does it mean that they are not crucial to ecosystems or that their disappearance will have no serious ramifications for people.

It certainly does not mean scientists ignore the jeopardy of overlooked species, even if the media do. A sampling of studies shows the wide range of research about threatened and endangered species: abalone in the Canary Islands (Riera et al., 2018); southern brown bandicoots in Australia (Maclagan, Coates, & Ritchie, 2018); Caspian sturgeon in Iran (Aghilinejhad et al., 2018); the Apennine chamois (Formenti et al., 2018) and quillwort (Abeli, Cauzzi, Rossi, Pistoja, & Mucciarelli, 2018) in Italy; dragonflies in Germany (Holtmann, Juchem, Brüggeshemke, Möhlmeyer, & Fartmann, 2018); the Eastern massasauga rattlesnake in the United States (Hileman, King, & Faust, 2018); Knysna seahorses in South Africa (Claassens & Hodgson, 2018); striped hyenas in Kenya (Ndeereh, Oloo, Rono, Manoa, & Bundotich, 2019); the wandering albatross of the Indian Ocean (Weimerskirch, 2018); and many, many more.

The role of environmental communication

Most people get the majority of their news and knowledge about the environment from the mass media. Conservation biologist David Schiffman and his colleagues note that:

> One important mechanism by which scientific information spreads across the science/policy interface is through media coverage from the popular press, which plays a role in public understanding and interpretation of scientific and environmental issues. Popular press coverage not only reflects current public opinion on a topic but has the ability to shape and influence current public opinion, with potentially large implications for public policy change.
>
> (2020, p. 3)

Beyond educating and informing the public, communication about biodiversity and other environmental issues informs policy making, from what lawmakers legislate to what administrators promulgate, from what politicians campaign about to what civil servants implement, and from what advocacy groups lobby for – or against – to what businesses and industry do.

Media effects research provides insights into the relationship between environmental concern and media use. "Media use keeps people informed of environmental issues that potentially encourage them to engage in environmental citizenship" (Takahashi, Tandoc, Duan, & Van Witsen, 2015, p. 128), which is defined as "the engagement in political activities aimed at supporting environmental causes" (p. 114).

Communication of cultural values and traditions within communities is also critical to biodiversity and species preservation. Thus messages and information get communicated from person to person and from small group to small group, often from parents to children, from grandparents to grandchildren, from neighbor to neighbor, from elder to younger. They include traditional ecological knowledge. As a study of kahuli, a species of endangered Hawaiian land snails, points out, "Centuries[-]old observations preserved within song, chant, and story ha[ve] been globally recognized as a resource to integrate with conservation efforts for endangered species" (Sato, Price, & Vaughan, 2018). To illustrate, researchers looking for rare orchids in Turkey relied on local residents for assistance. As their study observed:

> Local traditional communities often possess rich ecological knowledge connected to nature and traditional resource use and management. This knowledge can provide unexpected new information for researchers and

> show new opportunities and ways for professionals in conserving rare and threatened species.
>
> (Molnár, Süveges, Molnár, & Löki, 2017)

On the other hand, long-standing cultural traditions can contribute to sustainability problems. A study in the Republic of Georgia's Batumi Bottleneck area along the Black Sea coast found that poachers annually illegally kill thousands of migratory birds of prey, such as greater spotted eagles and pallid harriers. One major factor, the researchers found: "[T]he practice is often promoted as a part of hunting traditions and an important custom in the coastal villages" (Sándor & Anthony, 2018, p. 11851).

Our age of rapidly transforming and transformative communication technology makes it easier to disseminate factual, science-based environmental information, thus furthering the public policy goals of environmental literacy, consciousness, and action. Dissemination comes from the traditional or legacy news media that have historically played an essential role as agenda-setters (McCombs, 2005) and gatekeepers (Shoemaker & Vos, 2009). In addition, it comes increasingly from alternative and specialized news organizations, activists, industry, advertising, scientists, citizen scientists, government agencies, and ordinary social media users.

At the same time, the pervasiveness and ease of access to social media opens the door to abuses, including dissemination of "fake" or misleading environmental news and facilitation of illegal trafficking in threatened species.

As to the first type of abuse, investigative journalist Carey Gillam notes that "misinformation about relevant science has been amplified to polarize views" on climate change through such ubiquitous platforms as Facebook (2020, p. 83). But the plethora of environmental misinformation goes beyond climate, Gillam says, citing the fossil-fuel industry where "[t]he scientific arena is a hot spot for the manipulation of information" (p. 84). Shirley Ho and Silje Kristiansen observe, "Research has shown that fake news gets amplified and travels faster than positive news on Twitter," and "the widespread dissemination of fake news and misleading information on social media could also impact public discourse and perceptions about nuclear energy" (2019, p. 436). David B. Sachsman and JoAnn Myer Valenti put it more bluntly: "While some see the Internet as the future of environmental journalism, the web is a very different medium in which everyone can say whatever they wish. True journalists are in a minority among the hucksters and propagandists" (2020, p. 2). As for the implications of the proliferation of false "news" about environmental matters, Shiffman and colleagues say:

> When concerned non-experts wrongly believe that a problem is worse than it really is, or when they are misinformed about which policy solutions exist, they may be more likely to support more extreme solutions that are less supported by scientific evidence.
>
> (2020, p. 4)

The second type of abuse involves the misuse of social media in ways that exacerbate the threat to endangered species. Here are two examples: A Chilean journalist reported that smugglers use Instagram and eBay to illicitly peddle rare, slow-growing cacti from the coastal Atacama Desert (Almendras, 2019). Spee and colleagues reported how "the exotic pet trade [in the Middle East] is impacted by social media via greater accessibility to photos and videos including these species and the increasing popularity of online animal marketplaces. The social media presence of public figures owning exotic pets has a major influence on exotic species demand" (2019, p. 480). They found that 85% of species shown on social media posts are listed in the Convention on International Trade in Endangered Species of Wild Fauna and Flora (CITES) Appendices I and II, "including big cats, birds of prey, and great apes" (p. 480).

Overview of the book

Extinction isn't new – witness the evidence of Benjamin the thylacine in Australia, Martha the passenger pigeon in North America, and the anonymous Caspian tiger in Uzbekistan. But as the pace of extinction ramps up, the willingness and the ability of ordinary people, governments, scientists, nongovernmental organizations, and businesses to slow this deeply disturbing acceleration are uncertain. Meanwhile, researchers labor to better understand and communicate the possibility and implications of extinctions and to discover effective tools and public policies to combat the threats to species survival.

This multidisciplinary book takes a distinctive approach by connecting how media depict and explain endangered species with how policy makers and natural resource managers can or do respond to these challenges in practical terms. It draws on the latest research by social scientists, communication experts, natural resource managers, and conservation scientists whose work reflects the intersection of scientific inquiry about threatened and endangered species with communication and public policy. The interplay of science, culture, economics, policy, human emotion, resource management, and, of course, communication supports the book's three principal aims:

- Illuminate the interconnected nature of habitat, biodiversity, and species extirpation and extinction challenges around the world.
- Spotlight recent research in the social and natural sciences regarding the challenges faced by endangered species.
- Provide cultural, social, economic, and political contexts to recognize and tackle those challenges and problems.

The editors acknowledge the impossibility of covering all facets of communication and policy pertaining to endangered species in a single volume. That is why this book cannot include all relevant topics that are, nonetheless, subjects of fascinating research with potential practical impact on the biodiversity crisis. Omitted topics include the influence of extinction-related movies and television shows on public attitudes, the role of "habitat islands" such as cemeteries and sacred groves, and the ethical implications of cloning extinct or near-extinct species, among others.

The book organizes four closely connected themes under these headings: 1) News Coverage of Endangered Species Around the World; 2) Images of Endangered Species: Communication and Public Opinion; 3) Media Emphasize the Charismatic: Ignoring the Rest; and 4) Environmental Public Policy.

Bill Kovarik begins the section on News Coverage of Endangered Species Around the World by quoting Thomas Jefferson's belief that it was inconceivable that species could become extinct: "Such is the economy of nature," Jefferson said, "that no instance can be produced of her having permitted any one race of her animals to become extinct" (1787, p. 366). In "Endangered species, news, and public policy: A history," Kovarik tells the story of the development of our modern ideas of extinction and conservation from the 18th century to the current era. Beginning with the doomed dodo, Kovarik describes the decline of the buffalo, the "murder for millinery" (the slaughter of birds for hat plumes), the extinction of the passenger pigeon, and the rise of the Conservation Era. Kovarik argues that the "main elements of a social response found in news coverage involved scientific discussions, ethical movements and public policy debates" and that news of science, sometimes spiced with sensationalism, initiated the public reaction to endangered species and extinction.

Michael A. Patten and Brenda D. Smith in "The extinction crisis: Why words matter" argue that the media, the general public, and even conservation scientists often misuse the word "extinction," which has and *should* have a very specific meaning: that a species is dead and gone, now and forever, everywhere. Even scientists talk about "local extinction," when what they mean is "extirpation." But few journalists or members of the public know that *extirpation* means that an organism or species ceases to exist in a particular area, while the word *extinct* is known, used, and grabs attention, and therefore is often misused.

Much of the history of North American wolves involves their slaughter by hunters, ranchers, and farmers, and the wolf continues to be feared and hated by many. Alexander Killion, Mark Neuzil, and Eric Freedman report that "the wolf ranks among the world's 10 most charismatic animals, yet it is the only one the IUCN classifies as of 'least concern,' meaning it is not a focus of conservation efforts (Courchamp et al., 2018)." Nevertheless, "The wolves of fate: Media coverage of the Isle Royale 'genetic rescue'" describes the National Park Service (NPS) program to reintroduce gray wolves into a remote island park in Michigan. They analyzed 68 articles on the translocation and found that "journalists used large portions of the information from press releases, had no alternative sources for up-to-date photos and video, and heavily relied on the same sources proffered by the NPS and the Joint Information Center for additional quotes or data." They concluded that "by controlling access to locations where news occurs, authorities infringe on reportorial independence and the ability of journalists to gather information through the exercise of their own professional judgment."

Unlike the wolf, the elephant is a popular, charismatic animal, loved by almost everyone except the poachers who kill them for their ivory tusks and local farmers who fear them as a "detriment [to] Indigenous livelihoods." In "How African is the African elephant? Reflections on poaching and conservation in African and UK newspapers," Harriet Croome and Clare Wilkinson describe their study of 21 years of news coverage of the "elephant" and "poaching or conservation." They found that "representations differ across continents, with journalists in the UK and Africa employing different frames and sources and arranging them according to their perception of what matters to their respective home audiences." They argue that "the media's representation of elephant conservation may help perpetuate 'fortress'-style conservation initiatives that separate people from wildlife. More important, perhaps, is the media's potential to disengage local Africans from conservation by emphasizing the views of elite, often external actors."

In Part II: Images of endangered species: Communication and public opinion, Gavin Lamb's "Sea turtle tourism and conservation: Conflict and concord on a Hawaiian beach" tells the story of, and describes the images of, Hawaiian green sea turtles. This turtle is so popular that it became a major tourist attraction, which creates a conflict between this tourism and conservation efforts. "The influx of hundreds of thousands of tourists to a small rural community lays bare fissures and frictions about what the future landscape of human–sea turtle relations should look like in the state," Lamb reports. "Exploring communication at the nexus of sea turtle conservation and tourism in Hawai'i raises critical questions about the different degrees of power that conservation, tourism, and local community actors hold to influence people's experiences and interactions with threatened wildlife." He continues, "In Hawai'i, the dominance of the state-supported ecotourism industry increasingly drives alternative stories in its profit-driven direction, while framing itself as the solution to, rather than the cause of, problematic human-wildlife relations."

Monkeys and apes are as well-liked as turtles and elephants, and since 2000, the Primate Specialist Group of the IUCN's Species Survival Commission has kept a "Top 25 List" of the world's most endangered primates. Every two years, group members battle over which endangered primates will be given the worldwide attention engendered by inclusion on the list. In "The 25 most endangered primates list: Impact on conservation fundraising and policy," Kim E. Reuter et al. report that "the Top 25 List is the only media communications tool of its kind that highlights the plight of primates across the world." They continue:

> Within that context and looking back on its evolution, inclusion helped mobilize significant financial resources for many dozens of threatened species and has led to policy and governance changes in national hunting laws, government national action and species protection plans, and expansion or creation of protected areas.

Their analysis: "[A]s a communications tool, it has had significant impacts on donor and government awareness of, and support for, primate conservation worldwide."

Michael Scott Smith reports in "You better believe it: Rhino horns are a bloody risky business" that there are only 20,000–30,000 rhinoceroses remaining in the world. Rhinos are poached for their horns, which are used in traditional Chinese medicine in parts of Asia including Vietnam. Environmental nongovernment organizations sponsor efforts in end-market nations such as Vietnam to conduct behavioral change campaigns and encourage legal efforts against the illicit trade. Smith reports the results of his news framing study of "conservation groups' press releases on an aspect of the illegal wildlife trade in Vietnam." He found that Vietnamese ENGOs use four frames to demonstrate what is at stake: Responsibility; Lip Service; Voodoo Wildlife Parts; and Youth Empowerment. This chapter focuses on the Responsibility frame because a key challenge for ENGO campaigns has been to make Vietnamese consumers and the government see their role in driving the rhino poaching crisis (Smith, 2017) and the consequences of their actions.

Part III: "Media emphasize the charismatic: Ignoring the rest" begins with the most popular, the most beloved, the most charismatic animal in the world, the giant panda, which may have been the first animal in the world to be viewed every day in a live feed from its habitat in a zoo. In "Front pages are for the charismatic: The case of the cute giant panda," Ryan Wallace seeks to understand how some species, like the giant panda, have become "icons for nations, organizations, and environmental movements." He concludes that "giant pandas have proven to be prime ambassadors and fundraisers. They have had significant economic and sociocultural effects. Giant pandas have also facilitated change regarding some conservation efforts, prompting nations to notice the exorbitant efforts necessary to bring a species back from the edge of extinction." However, he notes that the emphasis on charismatic species like giant pandas "has raised questions regarding the implications of focusing on such flagship species," questions about the discourse of conservation and on policies and management strategies."

The critical environmental issue of the 21st century is climate change, and the icon of climate change is the polar bear. A charismatic animal to begin with, it is easy to explain and understand the effects of global warming on polar bears. A simple picture of a polar bear stranded on a small block of ice in the middle of vast ocean is all it takes to communicate the dilemma. Dorothea Born traces the public image of the polar bear in "Polar bears as cultural symbols: Threatening monsters and threatened species." As recently as 1985, "the documentary 'Polar Bear Alert' on National Geographic TV depicted them as bloodthirsty monsters." This change "'from fierce to adorable' (Archibald, 2015, p. 266) resulted from scientific advances in polar bear population monitoring and popular accounts of this research." Then, with the rise of climate change, the polar bear became its icon. Born illustrates this change by examining magazine articles about polar bears in both *National Geographic* and *GEO* from 1992 to 2012. She concludes that "the use of polar bear images to raise climate change awareness may actually inhibit a broader cognitive understanding of the issue (Manzo, 2010). The icon may attract attention and trigger emotional reactions, but it may not make the larger causes of nature exploitation and climate change more visible (Hansen & Machin, 2013)."

Social media, like traditional media, are often "biased toward charismatic or eye-catching species like wolves, American bison, eagles, and bears (Roberge, 2014)," report Caleb R. Hickman and Shem D. Unger in "Societal attitudes towards hellbender salamander conservation: The roles of traditional and social media." "But how would the public respond to less charismatic and less-easily observed species? One such poorly recognized species is the [fully aquatic] hellbender salamander," which has "a tumultuous history of conservation, persecution, endearment, and public fear." The authors evaluated print and online media representations of the hellbender salamander to determine whether "articles were relatively positive, negative, or neutral in nature." While hellbenders had historically received negative perceptions, Hickman and Unger found that public attitudes have been positive since 1980 in both print and online media, and that social media in particular "have seen a large increase in coverage of hellbenders, with an overall increase in engagement, as well as more scientific research about this enigmatic animal." The hellbender salamander, once considered ugly, is now becoming popular.

How have news reports of endangered species affected public policy in nations around the world? This is the fundamental research question of *Communicating Endangered Species: Extinction, News and Public Policy*. While all the chapters address this question in one way or another, the four chapters in Part IV: "Environmental public policy" directly address this issue.

In "(Not) saving the Great Barrier Reef from disaster: Media then and now," Kerrie Foxwell-Norton and Libby Lester describe efforts of the "Save the Reef" campaign in Australia in the 1960s and 1970s to protect the Great Barrier Reef from mining and oil and gas exploration. "In September 1969, *The Courier Mail* published a map of the reef region leased for oil drilling, naming the companies with leases – details that had been shrouded in secrecy until then." With 80% of the Reef leased for oil drilling, federal politicians raised a "parliamentary enquiry that eventually led to passage of the Great Barrier Reef Marine Park Act," which established "Commonwealth rather than state jurisdiction over management of the Reef."

Foxwell-Norton and Lester then turn to the "Fight for the Reef" campaign, which began in 2012. While still concerned with "local environmental policies designed to administer and protect the Reef from destruction," the current campaign's "communication strategy is aimed at transcending national boundaries and bypassing traditional channels of power associated with media, government, and industry." They continue:

> To quantify, in comparison to the 13,000 signatures delivered to the [Australian] Parliament in 1968, the Fight for the Reef Facebook site had over

> 170,000 likes, and its online petition with 81,000 signatures was delivered to the World Heritage Committee Meeting in Cambodia in 2013.

Do celebrity media campaigns affect environmental public policy? Elaine Jeffreys, in "Communicating shark extinction: Celebrity-endorsed conservation campaigns and public policy in China," explains that "supporters argue that leveraging 'fame' raises public awareness of environmental issues and organizations, as well as highlighting the need for legislative change by generating media publicity, clarifying campaign issues, and attracting new audiences and donors (https://wildaid.org). Critics counter that celebrity-fronted environmentalism is a self-serving celebrity branding exercise that oversimplifies issues by providing 'spectacular enviro-tainment.'"

Like wolves, sharks are feared, but also like wolves, they are charismatic. Some 100 million sharks are killed annually, mainly for shark-fin soup, a Chinese delicacy. "Overfishing threatens to destroy the balance of ocean marine life by placing multiple shark species at risk of extinction," explains WildAid, an international conservation organization that "has used celebrities such as Shanghai-born Yao Ming, formerly a star player with the North American National Basketball Association, as the public face of communication aimed at ending the consumption of shark-fin soup in Chinese restaurants globally." Jeffreys conducted a media content analysis using "combinations of the keyword 'shark fin' with, respectively, 'Yao Ming,' 'WildAid,' 'public money,' and 'corruption.'" She found "the media response to WildAid's conservation communication was greater in international arenas than in China," continuing: "Journalists credit the Yao Ming PSAs [public service announcements] for contributing to bans on shark fin in 2011 in [the state of] California and [the Canadian city of] Toronto – areas with large proportions of ethnically Chinese residents (Harmon, 2011; Rogers, 2013)." She concludes: "Although the decline in China's shark-fin trade cannot solely be attributed to celebrity-endorsed conservation communication and government austerity measures, both contributed to a reduced demand" for shark fin.

John Carr and Tema Milstein in "Manatees and fossil-fuel power plants: The paradox of endangered species laws" describe the strange, ironic situation in which Florida in 1979 adopted manatee protection zones that include power plant effluent, and in which "in 1999, the three agencies charged with protecting manatee habitat under state and federal law – the Florida Fish and Wildlife Conservation Commission, the US Fish and Wildlife Service, and the Florida Department of Environmental Protection – approved the first Power Plant Manatee Protection Plans." Thus the large, aquatic, mostly herbivorous marine mammal known as the sea cow, which seeks the warmth of power plant effluent, protects the future – and even requires the future – of coal and hydrocarbon gas–burning power plants, facilities that environmentalists elsewhere campaign against. They conclude that "by positing fossil-fuel-fired power plants as part of a network of endangered species sanctuaries, the regulatory regime . . . obscures their destructive environmental impacts."

"When a species is listed as endangered or threatened under the Endangered Species Act (ESA), restrictions begin on uses of both public and private land," explain Joe Kerkvliet, Christian Langpap, and Jason Shogren in "Conservation on private land: The Endangered Species Act." They report that "there is substantial evidence that ESA regulations or the threat of regulations have induced some landowners to pre-emptively modify habitat to discourage imperiled species from occupying it (Langpap, Kerkvliet, & Shogren, 2018)." They fear that "existing regulations may actually have slowed progress towards achieving environmental goals and caused voters to voice opposition to regulations and/or taxes. Economic and political opposition has been strong and often threatens the strength, and even the existence, of environmental laws." The authors conclude that "the purposes of the ESA are unlikely to be achieved without actively engaging the resources of private land and the support of private landowners."

In the final chapter, "Preserving endangered species: Communicating, reporting, and framing the extinction crisis," the editors discuss translocations and reintroductions of endangered species, the role of zoos and aquariums, news coverage and framing, and the use of celebrities in conservation campaigns. At the end, they look into a very cloudy crystal ball and speculate about key factors and trends that may influence the future of environmental communication about endangered species and its impact on public policy.

Several threads connect these chapters, much like the interconnected habitats of threatened and endangered flora and fauna connect with their physical relationship to human habitation and activities. The first thread is the impossibility of separating science, policy, and communication if public and private institutions and individuals are to take effective measures to avert species extinction. Field and laboratory research findings, as well as social science research findings, should inform decision-making by governments and nongovernmental organizations. Communicating the results of that research should inform, educate, and motivate the citizenry and influence how people act individually and collectively to address the biodiversity crisis.

A second commonality is the ineffectiveness of physical and political borders to define problems of prospective extinctions and to identify possible solutions. By definition, invasive species pay no heed to such borders. Nor do natural disasters, such as tsunamis and hurricanes. Nor does climate change. Nor do plant pests and animal diseases. Nor do toxic spills. Nor do travel paths of airborne pollutants and plumes of contaminated groundwater. At the same time, borders do matter because of their political and diplomatic significance. Which public agencies will take remedial, preventative, legislative, and regulatory actions? Who will pay for such measures? How should the economic burden to protect biodiversity be shared among wealthy and poor countries alike?

A third thread is the media's responsibility to present information and fact-based opinion about threatened species and biodiversity in a fair, accurate, balanced, and ethical way. This includes avoidance of the-sky-is-falling-and-there-is-no-place-to-hide over-sensationalizing, as well as avoidance of facilely minimizing true risks. A priority is public credibility. Journalists and other communicators should greet with skepticism "problems" or "solutions" proffered without scientific evidence. They should ask tough questions of scientists, advocacy groups, politicians, economists, natural resources managers, and government officials. Additionally, when the topic is extinction, their reporting should go beyond rhinos and polar bears, wolves and manatees, and other megafauna and charismatic species that typically garner the most media attention. Journalists and other communicators should offer context and point out the mutual dependencies of endangered and non-endangered species.

Experts disagree on the number of species that go extinct on average every day, with science-grounded estimates ranging from 24 to 150. The precise number is unimportant, however. Endlings simply disappear, almost always anonymously and unnoticed. Here one day. Gone the next. Then another disappears. Then another.

Most endlings won't be as "lucky" as Benjamin the thylacine, Martha the passenger pigeon, and the unnamed Caspian tiger mounted in an out-of-the way Central Asian museum. They won't get a name, a numbered box, or a dusty diorama in a museum, or a plaque at a zoo. In most cases, we're unlikely to even identify them before it's too late, let alone name them.

The question is: What lessons can their plight teach us as communicators, journalists, natural scientists, social scientists, policy makers, resource managers, and environmentally concerned individuals? The editors hope this book assists in the search for meaningful and pragmatic answers to that question.

Acknowledgments

This chapter draws on material previously published in Freedman, E. (2011). Extinction is forever: A quest for the last known survivors. *Earth Island Journal* (Autumn) and is published with permission from *Earth Island Journal*. It also draws from material previously published in Freedman, E. (2002, Winter). Cut from history. *EJ Magazine*, 1–18, and is published with permission from the Knight Center for Environmental Journalism, Michigan State University.

References

Abeli, T., Cauzzi, P., Rossi, G., Pistoja, F., & Mucciarelli, M. (2018). A gleam of hope for the critically endangered *Isoetes malinverniana*: Use of small-scale translocations to guide conservation planning. *Aquatic Conservation, 28*, 501–505.

Aghilinejhad, S. M., Gorgin, S., van Uhm, D., Joolaie, R., Ghorbani, R., Paighambari, S. Y., . . . Jalali, A. (2018). What are the drivers of the occurrence of illegal fishing and conservation barriers of sturgeons in the Caspian Sea? *Aquatic Conservation, 28*, 690–701.

Almendras, D. (2019). Modern smugglers use social media to sell Chile's ancient botanical riches. Knight Center for Environmental Journalism. Retrieved from https://knightcenter.jrn.msu.edu/2019/01/04/modern-smugglers-use-social-media-to-sell-chiles-ancient-botanical-riches/

Archibald, K. (2015). From fierce to adorable: Representations of polar bears in the popular imagination. *American Review of Canadian Studies, 45*(3), 266–282.

Campbell, C. R. (n.d.). *The thylacine museum*. Retrieved from www.naturalworlds.org/thylacine/

Ceballos, G., Ehrlich, P. R., & Raven, P. H. (2020). Vertebrates on the brink as indicators of biological annihilation and the sixth mass extinction. *Proceedings of the National Academy of Sciences, 117*(24).

Claassens, L., & Hodgson, A. N. (2018). Monthly population density and structure patterns of an endangered seahorse *Hipppocampus capensis*: A comparison between natural and artificial habitats. *Journal of Fish Biology, 92*, 2000–2015.

Courchamp, F., Jaric, I., Albert, C., Meinard, Y., Ripple, W. J., & Chapron, G. (2018). The paradoxical extinction of the most charismatic animals. *PLoS Biology, 16*(4), e2003997.

Dickie, G. (2020, July 20). Most polar bears to disappear by 2100, study predicts. *The Guardian*. Retrieved from www.theguardian.com/environment/2020/jul/20/most-polar-bears-to-disappear-by-2100-study-predicts-aoe

Fears, D. (2020, July 9). The North Atlantic right whale is the face of an extinction crisis, report says. *Washington Post*. Retrieved from www.washingtonpost.com/climate-environment/2020/07/09/north-atlantic-right-whale-is-face-an-extinction-crisis-report-says/

Formenti, N., Viganó, R., Fraquelli, C., Trogu, T., Bonfanti, M., Lanfranchi, P., . . . Ferrari, N. (2018). Increased hormonal stress response of Apennine chamois induced by interspecific interactions and anthropogenic disturbance. *European Journal of Wildlife Research, 64*, 68. doi:10.1007/s10344-018-1228-4

Fuller, E. (2014). *The passenger pigeon*. Princeton, NJ: Princeton University Press.

Gillam, C. (2020). Finding and following the facts in an era of fake news. In D. B. Sachsman & J. M. Valenti (Eds.), *Routledge handbook of environmental journalism* (pp. 83–94). London and New York: Routledge.

Hansen, A., & Machin, D. (2013). Researching visual environmental communication. *Environmental Communication*, 7(2), 151–168.

Harmon, S. (2011, October 7). Shark trade banned in California: Gov. Jerry Brown signs bill banning shark fins in California. *Mercury News*. Retrieved from www.mercurynews.com/2011/10/07/gov-jerry-brown-signs-bill-banning-shark-fins-in-california/

Hileman, E. T., King, R. B., & Faust, L. J. (2018). Eastern massasauga demography and extinction risk under prescribed-fire scenarios. *Journal of Wildlife Management, 82*(5), 977–990.

Ho, S. S., & Kristiansen, S. (2019). Environmental debates over nuclear energy: Media, communication, and the public. *Environmental Communication, 13*(4), 431–439.

Holtmann, L., Juchem, M., Brüggeshemke, J., Möhlmeyer, A., & Fartmann, T. (2018). Stormwater ponds promote dragonfly (*Odonata*) species richness and density in urban areas. *Ecological Engineering, 118*, 1–11.

Intergovernmental Science-Policy Platform on Biodiversity and Ecosystem Services. (2019). *UN report: Nature's dangerous decline "unprecedented"; Species extinction rates "accelerating."* United Nations. Retrieved from www.un.org/sustainabledevelopment/blog/2019/05/nature-decline-unprecedented-report/

International Union for Conservation of Nature. (2019a). *A miniature world in decline: European Red List of mosses, liverworts, and hornworts*. Retrieved from https://portals.iucn.org/library/sites/library/files/documents/RL-4-027-En.pdf

International Union for Conservation of Nature. (2019b). *European Red List of trees*. Retrieved from https://portals.iucn.org/library/sites/library/files/documents/RL-4-026-En.pdf

International Union for Conservation of Nature. (2020). *The IUCN Red List in a nutshell.* Retrieved from www.iucn.org/resources/conservation-tools/iucn-red-list-threatened-species

Jacobs, J. (2019, January 10). George the snail, believed to be the last of his species, dies at 14 in Hawaii. *New York Times.* Retrieved from www.nytimes.com/2019/01/10/science/snail-dead-george-species.html?searchResultPosition=1

Jefferson, T. (1787, April). Comparative view of the animals of America and those of Europe. *Columbian Magazine.*

Langpap, C., Kerkvliet, J., & Shogren, J. (2018). The economics of the U.S. Endangered Species Act: A review of recent developments. *Review of Environmental Economics and Policy, 12*(1), 69–91.

Maclagan, S. J., Coates, T., & Ritchie, E. G. (2018). Don't judge habitat on its novelty: Assessing the value of novel habitats for an endangered mammal in a peri-urban landscape. *Biological Conservation, 223*, 11–18.

Manzo, K. (2010). Beyond polar bears? Re-envisioning climate change. *Meteorological Applications, 17*(2), 196–208.

McCombs, M. (2005). A look at agenda-setting: Past, present and future. *Journalism Studies, 6*, 543–557.

Molnár, A. V., Süveges, K., Molnár, Z., & Löki, V. (2017). Using traditional ecological knowledge in discovery of rare plants: A case study from Turkey. *Acta Societatis Botanicorum Poloniae, 86*(3), 3541.

Mongabay.com. (2019, April 16). Last known female Yangtze giant softshell turtle dies in China. Retrieved from https://news.mongabay.com/2019/04/last-known-female-yangtze-giant-softshell-turtle-dies-in-china/

Ndeereh, D., Oloo, T., Rono, B., Manoa, D., & Bundotich, G. (2019). Towards saving an endangered species: Rehabilitation of hand-raised striped hyena cubs in Meru National Park, Kenya. *African Journal of Ecology, 57*, 144–147.

Rice, D. (2019, January 2). These species went extinct in 2018. More may be doomed to follow in 2019. *USA Today.* Retrieved from www.usatoday.com/story/news/nation/2018/12/31/extinct-species-these-animals-were-lost-forever-2018/2450121002/

Rice, D. (2020, February 7). Bumblebees are disappearing at rates "consistent with mass extinction." *USA Today.* Retrieved from www.usatoday.com/story/news/nation/2020/02/06/bumblebees-decline-due-climate-change/4679240002/?_ga=2.8211376.32081612.1596652234-1988510857.1596652234

Riera, R., Herrera, R., Pérez, Ó., Garrido, M. J., Álvarez, Ó., Monterroso, O., & Nuñez, J. (2018). Lack of recovery symptoms of an endangered and harvested mollusc in the last 20 years. *Journal of the Marine Biological Association of the United Kingdom, 98*(2), 351–355.

Roberge, J. (2014). Using data from online social networks in conservation science: Which species engage people the most on Twitter? *Biodiversity and Conservation, 23*, 715–726.

Rogers, P. (2013, July 1). Shark fin soup ban takes effect Monday. *Mercury News.* Retrieved from www.mercurynews.com/science/ci_23567375/shark-fin-soup-ban-takes-effect-monday

Sachsman, D. B., & Valenti, J. M. (2020). Introduction: Environmental journalism. In D. B. Sachsman & J. M. Valenti (Eds.), *Routledge handbook of environmental journalism* (pp. 1–15). London and New York: Routledge.

Salafsky, N., Salzer, D., Stattersfield, A. J., Hilton-Taylor, C., Neugarten, R., Butchart, S. H. M., . . . Wilkie, D. (2008). A standard lexicon for biodiversity conservation: Unified classification of threats and actions. *Conservation Biology, 22*, 897–911.

Sándor, A., & Anthony, B. P. (2018). Mapping the conflict of raptor conservation and recreational shooting in the Batumi Bottleneck, Republic of Georgia. *Journal of Threatened Taxa, 10*(7), 11850–11862.

Sato, A. Y., Price, M. R., & Vaughan, M. B. (2018). Kāhuli: Uncovering Indigenous ecological knowledge to conserve endangered Hawaiian land snails. *Society & Natural Resources, 31*(3), 320–334.

Shiffman, D. S., Bittick, S. J., Cashion, M. S., Colla, S. R., Coristine, L. E., Derrick, D. H., . . . Dulvy, N. K. (2020). Inaccurate and biased global media coverage underlies public misunderstanding of shark conservation threats and solutions. *iScience, 23*, 101205.

Shoemaker, P., & Vos, T. (2009). *Gatekeeping theory*. New York, NY: Routledge.

Smith, M. S. (2017). *The framing of rhino horn demand reduction by Vietnamese ENGOs: Cultural and other factors influencing the "responsibility" frame in ENGO media outputs* (Master's thesis). Griffith Research Repository. Retrieved from http://hdl.handle.net/10072/371949

Spee, L. B., Hazel, S. J., Grande, E. D., Boardman, W. S. J., & Chaber, A. L. (2019). Endangered exotic pets on social media in the Middle East: Presence and impact. *Animals, 9*(8), 480. doi:10.3390/ani9080480

Takahashi, B., Tandoc, E. C., Duan, R., & Van Witsen, A. (2015). Revisiting environmental citizenship: The role of information capital and media use. *Environment and Behavior, 49*(2), 111–135.

Weimerskirch, H. (2018). Linking demographic processes and foraging ecology in wandering albatross –conservation implications. *Journal of Animal Ecology, 87*(4). doi:10.1111/1365-2656.12817

Yong, E. (2019, February 19). Is the insect apocalypse really upon us? *The Atlantic*. Retrieved from www.theatlantic.com/science/archive/2019/02/insect-apocalypse-really-upon-us/583018/

Part I

News coverage of endangered species around the world

2 Endangered species, news, and public policy

A history

Bill Kovarik

Public concern about endangered species dates back to the early 19th century, when the very idea that a species could become extinct was a challenge for a society based on concepts of biblical creation and Aristotelian stasis.

This chapter traces the history of 19th- and 20th-century media responses to news about extinction, endangered species, and the rapid decline of natural systems. It concludes that the main elements of a social response found in news coverage involved scientific discussions, ethical movements, and public policy debates. These all played their respective parts in mediating responses to extinction and near-extinctions.

For example, 19th-century responses to rapidly declining numbers of bison led to the rise of biological surveys, then hunter-based conservation ethics movements, and then government-sponsored wildlife reserves. Similarly, ornithologists organized by the Progressive movement advanced early-20th-century tropical bird protection that was then supported by state and federal laws against bird poaching. The role of the US news media in the Progressive Era (1890s–1920s) was to report and encourage each phase of the scientific, ethical, and legal response, often against social inertia but without much controversy.

However, by the early to mid-20th century, complications and conflict emerged with wildlife conservation issues, especially over federal spending. To promote reform during the New Deal era, popular journalists joined government agencies to organize agencies and advocate conservation. By the 1950s and 1960s, journalists saw environmental concerns as "the great awakening." As the environmental movement took on new urgency, the news media moved beyond comfortable conservation columns into highly contested areas of issue advocacy and challenges to institutional architecture.

In the 1970s, in the aftermath of Earth Day and the popular embrace of environmentalism, many news organizations began to assign reporters to permanent environmental beats. But existential questions pointed up fundamental flaws in the democratic theory of neutrally reported, factually based news coverage: What limits should be placed on the environmental agenda and how can viewpoints be balanced, given that the next extinct species could be humanity itself?

As the volume and political impact of environmental news coverage expanded in the late 20th century, higher stakes triggered more antagonistic responses from government and industry and scorched earth tactics from right-wing political movements. While the news media sometimes served as the fulcrum between polluting industries and government regulation, a frequent industry tactic was to escalate grassroots conservative activism, undercutting the balance of environmental discourse and leaving the fact-oriented news media struggling to be heard over the din of social media.

Breaking the great chain

The idea that a species could become extinct was inconceivable a few hundred years ago, contradicting the long-standing biblical and Aristotelian world view of a static and permanent order of creation known as the "Great Chain of Being" (Glacken, 1976, p. 481). For example, when future US President Thomas Jefferson argued with leading French naturalist Comte de Buffon in 1787 over fossils and the alleged "degeneracy" of North American mammals, they disagreed about the size and soundness of their respective great mammals, but they were in perfect agreement about their roles in the Great Chain of Being. "Such is the economy of nature," wrote Jefferson, "that no instance can be produced of her having permitted any one race of her animals to become extinct" (p. 366).

Jefferson and most others of his era resisted the concept of extinction despite mounting evidence being compiled by another French naturalist, Georges Cuvier (Rosenberg, 2009). In a 1796 lecture, Cuvier demonstrated that unique fossils found in different geological strata around Paris could be nothing other than *espèces perdues* – or lost species. Cuvier analyzed thousands of fossils, along with specimens of living animals, and developed a classification system for his 1817 encyclopedic *Animal Kingdom*. His 1832 obituary in the *Times of London* noted that the book was a major influence and that he "shook to the base" (p. 5) the zoological classifications which then prevailed. When British scientists published an expanded edition of *Animal Kingdom* in 1835 (p. 3), the *Times* lauded it as a major improvement over Swedish zoologist Carl Linnaeus's 1758 binomial classification *Systema Naturae*.

Scientists were not considered public figures at this time, and even sensational reports were rarely covered in the news. Richard Owen's 1841 (p. 103) report on the remarkable fossils that he labeled "dinosaurs" was ignored by the *Times of London*. Later, his dioramas of prehistoric dinosaurs, presented in London and New York in the 1850s–1870s, became popular as illustrations of changing ideas about science. But the initial accumulation of scientific facts about extinction was not the key to a profound shift in popular sentiment. That honor goes, instead, to a sensational bestselling book published in 1844, *Vestiges of the Natural History of Creation*, which broke the theological concept of the Great Chain of Being. Written in a popular style by an anonymous author, it reportedly held Victorian-era readers spellbound, presenting new ideas about the ancient origins of life as proven by geological strata and fossil finds (Secord, 2000).

"*Vestiges* attracted attention for the boldness with which it attacked accepted theories and conclusions," said the *Edinburgh Scotsman* in 1884, 40 years after its publication. "Immediately the bigots of science and of theology rose up in anger . . . and . . . pronounced on the book a greater malediction" (*New York Times*, 1884a, p. 3). But science was moving on, and species extinction had become a foundation of evolutionary theory. By the 1880s, scientists and most of the public accepted the idea of extinction, if not the entirety of Charles Darwin's theory of evolution.

Darwin's champion, Alfred Russel Wallace, is credited in the *Oxford English Dictionary* as the first to give a biological meaning for the word "extinct." "The most effective agent in the extinction of species is the pressure of other species," the dictionary quoted him as saying (Wallace, 1880, p. 16). Similarly, newspapers reflected acceptance of the idea around the same time. "Entire races of animals have disappeared in recent times" said a *New York Times* (1882, p. 6) article. Claims that wooly mammoths had been captured alive, which might not have raised a Jeffersonian eyebrow only a century beforehand, were now greeted with bemusement. If it's true, the *New York Times* said, that a hunter had captured a living mammoth as he claimed, it would be "an animal wonder before whose brilliant refulgence even the white elephant must turn pale" (1884b, p. 8).

Why a dodo?

There is nothing funny about species extinction, unless it involves a flightless, hapless bird with small wings and a comically quizzical expression on its face. Linnaeus himself implicitly acknowledged the humorous nature of the dodo by calling it *Didus ineptus* in the 1766 edition of his book *Systema Naturae* (Fuller, 2002). In 1831, the dodo was said to "afford us an example of the extinction of an animal in comparatively recent times" (De la Beche, 1831, p. 154). Nor was there anything humorous about a sober and ponderous 1848 ornithological treatise, "The Dodo and its Kindred," by Hugh Edwin Strickland and G.G. Melville. "It appears highly probable that death is a law of nature in the species as well as the individual," Strickland said (Strickland & Melville, 1848).

The following year, the *Times of London* appeared to be the first to employ the dodo in a humorous figure of speech. In an article about water pollution, the newspaper complained that Thames River salmon are "as extinct about Blackwall as the dodo in Madagascar" (1848, p. 4). The newspaper continued to use the dodo to great effect in following years, for example, to explain that Spanish colonialism still exists and is "no questionable dodo" (1853, p. 4).

This unusual combination of science and humor attracted so much attention that when the Oxford University Museum of Natural History opened in 1860, a dodo from its collection was one of the most popular exhibits. The dodo's reputation as the archetype of extinction probably owed less to science and more to Oxford professor Charles Dodgson's whimsical book, written for a friend's daughter named Alice. Dodgson, better known by his pen name Lewis Carroll (1865), described the dodo proposing a "caucus race" after swimming in a pool of Alice's tears in *Alice in Wonderland.*

Use of the dodo as a figure of speech continued in Britain with examples such as "Dodo race of real unmitigated Toryism" (Carr, 1874); a style of dress as rare as a dodo in rural France (*Times of London*, 1875, p. 12); and "The old dodo at Scotland Yard" (*Pall Mall Gazette*, 1886, p. 1).

The dodo first landed as a figure of speech on North American shores in a *New York Times* article (1869, p. 7) observing that good acting was "as rare as a white elephant or a dodo." It was also used to illuminate the pending loss of "hospitable old gentlemen" (*New York Times*, 1875, p. 4); the "old fashioned tramp" (*Washington Post*, 1889, p. 10); a "political dodo – Senator [John] Jones of Nevada" (*Washington Post*, 1878, p. 2); the traveling circus (*New York Times*, 1888, p. 16); and many more.

The figure of speech didn't go extinct after the turn of the 20th century. Novelist F. Scott Fitzgerald described dumb love with this metaphor: "Bridges had been a dodo about some YMCA man" (Turnbull, 1964). The term also was emblematic about disappearing organ grinders and monkeys (*New York Times*, 1924, p. X8); the National Recovery Administration's blue eagle symbol (Bruner, 1935, p. 1); the Missouri mule (*Washington Post*, 1951, p. B4); a book named *Such Darling Dodos* (Wilson, 1950); the "vanishing" American male (Dixon, 1962, p. A15); the movie theater newsreel (Shepard, 1967, p. 37); slow pitch softball (O'Malley, 1980, p. D12); and feminist heroines (Caryn, 1989, p. H15).

Clearly, the dodo proved useful as a humorous figure of speech, perhaps symbolizing the decline of older ideas about natural stability, or a reaction to the rising encroachment of science and technology in everyday life. In any event, the dodo, as simile and metaphor, lives on centuries past its physical extinction.

Ethical awakening

Aside from scientific discoveries and light humor, one of the most important influences on public discourse about extinction and other species involved a new ethical perspective about animal life. One starting point can be located in the lectures of Immanuel Kant, who said that a person was duty-bound to practice kindness to animals. "If he is not to stifle his human feelings, he must practice kindness towards animals, for he who is cruel to animals becomes hard also in his dealings with men" (1784).

This and similar perspectives found an institutional home with the Royal Society for the Prevention of Cruelty to Animals, established in London in 1824 (*Times of London*, 1824, June 17). The society existed to address the "many instances of horrid wanton cruelty which, thanks to the public press, do appear before the public," said an 1828 (p. 3) *Times of London* letter calling for support for the RSPCA and signed "Humanitas."

Around the same time, sentimental articles about animal intelligence began appearing in newspapers, such as an account of a dog rescuing another dog that was drowning (*Niles Register*, 1817) and a dog's barking that saved a ship from a reef (*Niles Register*, 1836). By the mid-19th century, the idea that animals had feelings and some intelligence became common in newspapers and magazines (*New York Times*, 1874, p. 2).

Around this same time, a broader reverence for life on earth emerged through the Romantic movement in literature and art. In American literature, authors such as Washington Irving, James Fennimore Cooper, Emily Dickinson, Ralph Waldo Emerson, and Henry David Thoreau wrote from a spiritual perspective that they found through observing the natural world. These were not always sympathetic to animal life. In *Moby Dick* (1851), Herman Melville described the "malevolent intelligence" of whales and said that whalers "can hardly regard any creatures of the deep with the same feelings that you do those of the shore." In contrast, Alexander Starbuck wrote with empathy for whales in *Harper's* magazine. "Imagination cannot conceive anything more awful than the butchery that now takes place. Terrified, the whale plunges from wave to wave" (1856, p. 466).

The strongest early expression of empathy for animals in popular literature involved a novel in the voice of a mistreated horse. Anna Sewell's best-selling 1877 novel *Black Beauty* tied religion to ethical treatment of animals. "Cruelty was the devil's own trade-mark," she wrote in the horse's voice, "and if we saw anyone who took pleasure in cruelty we might know who he belonged to, for the devil was a murderer from the beginning, and a tormentor to the end. On the other hand, where we saw people who loved their neighbors, and were kind to man and beast, we might know that was God's mark."

A similar distinction between cruelty and compassion emerges in Sarah Orne Jewett's 1886 short story, "A White Heron," in which the heroine expresses "sorrow at the sharp report of [a hunter's] gun and the sight of thrushes and sparrows dropping silent to the ground, their songs hushed and their pretty feathers stained and wet with blood. Were the birds better friends than their hunter might have been, – who can tell? Whatever treasures were lost to her, woodlands and summer-time, remember!"

Many other works of fiction, including *Wind in the Willows*, *White Fang*, *Jungle Book*, *Doctor Doolittle*, and Mark Twain's "A Dog's Tale," were published in the late 19th and early 20th century, allowing readers to achieve higher levels of empathy with the natural world. Through romantic and ethical perspectives, the door was also opened to non-fiction nature writing by authors such as John Muir, John Burroughs, Florence Merriam Bailey, and Edward Howe Forbush.

Science occasionally came into conflict with the animal ethics movement, especially in Britain, around questions concerning the cruelty of vivisection experiments on living animals. Between 1903 and 1910, opponents of these experiments (called "doggers") and medical students ("anti-doggers") fought in street riots, in the courts, and around symbols such as statues of dogs (Rupke, 1987).

During this period, American hunters also began to embrace a new kind of ethic that included preservation and habitat protection. One example was US President Teddy Roosevelt's refusal to kill a bear cub in an incident covered by newspapers at the time and depicted in a political cartoon (*Washington Post*, 1902, p. 1). The incident inspired the much-loved stuffed toy, the Teddy Bear.

This idea of an expanding ethical understanding of the natural world also served as the main theme of Aldo Leopold's 1948 classic book, *A Land Ethic*. The origin of ethics, Leopold said, involved relationships among individuals who were guided by great moral codes such as the Ten Commandments. An expanded ethical understanding involved relationships among individuals and society, exemplified by the Golden Rule. Now, an ethic was needed to deal "with man's relation to land and to the animals and plants which grow upon it" (Leopold, 1949). The trend line for the news articles that described efforts to protect the natural world in the late 19th and 20th century follows this expanding circle of ethical understanding.

Trail of the bison

The concept of extinction was first regarded with curiosity and mild humor, but the American public and press became alarmed at the rapid loss of the vast herds of bison in the late 19th century, and this aroused public opinion saved the remaining herds in the early 20th century.

Originally, the bison were the great icons of the frontier and a background presence in early American literature for authors like Washington Irving and James Fennimore Cooper. There were so many that it seemed impossible for them to go extinct. "At any time [before] 1836, a traveler might start at any point north or south in the Rocky Mountain range. . .[and] his road would always be among large bands of buffalo, which would never be out of view," said explorer John C. Frémont (*Niles Register*, 1845). But by 1845, the buffalo started to occupy "a very limited space," he wrote. "The extraordinary rapidity with which the buffalo is disappearing from our territories will not appear surprising when we remember the great scale on which their destruction is yearly carried on."

New York Tribune editor Horace Greeley was able to see the last of the great herds on his 1859 "Overland Journey" to California. "All day yesterday, they darkened the earth around us, often seeming to be drawn up like an army in battle array on the ridges and down their slopes a mile or so south of us," Greeley wrote on May 31, 1859. "What strikes the stranger with most amazement is their immense numbers. I know a million is a great many, but I am confident we saw that number yesterday" (Greeley, 1860, p. 87).

Like Greeley, young future president Roosevelt (1893) also traveled west and wrote of his bison hunts. "Their numbers were countless – incredible," he wrote.

> In vast herds of hundreds of thousands of individuals, they roamed from the Saskatchewan to the Rio Grande and westward to the Rocky Mountains.

> After the ending of the Civil War, the work of constructing trans-continental railway lines was pushed forward with the utmost vigor. Several million buffaloes were slain. In 15 years from the time the destruction fairly began, the great herds were exterminated.

Roosevelt's concern about the bison exemplified a new ethic of sportsmanship championed by a new breed of "sporting journals" led by *Forest and Stream*, established in 1873 by Charles Hallock and edited by George Bird Grinnell (Neuzil & Kovarik, 1996). Hallock helped form the New York Association for the Protection of Game in 1874, while Grinnell helped spark state Audubon societies and the Boone and Crockett club for hunters and conservationists.

A few people contradicted the calls for conservation, but not many. In one case, the superintendent of Yellowstone National Park insisted the herd of 200 bison was well protected and would not go extinct (*Washington Post*, 1888, p. 2).

But by the 1890s, conservationists were sounding dire warnings not only in specialized publications and sporting journals but also in mainstream newspapers. When William T. Hornaday, the superintendent of the National Zoo in Washington, DC, wrote "The Extermination of the American Bison," his subsequent lectures and ideas were carried in newspapers around the country (*Washington Post*, 1890, p. 8). Hornaday remained a leading scientific voice for saving bison and other endangered species through the early part of the 20th century. He was widely quoted in the press, for example, decrying the "unprecedented slaughter" of the bison in the *New York Times* (1896, p. 20). The alarm escalated towards the turn of the century. Bison extermination was "The Crime of the Century" according to an 1899 *Scientific American* article by Charles Frederick Holder, a naturalist and author. It is "a pitiful story of the greed of the white man and the extirpation of a mighty race within three decades" (p. 378).

Articles about the beasts quadrupled between the 1870s and 1880s, then doubled again in the 1890s. At the height of the Progressive Era's interest in bison, more than 9,000 articles ran in the *Washington Post*, *New York Times*, *Baltimore Sun*, and *Wall Street Journal* between 1890 and 1900, while another 20,000 ran between 1901 and 1910. Magazine writers such as Robert Underwood Johnson of *Century Magazine* warned that what happened to the bison could happen elsewhere, reflecting an enormous increase in interest about endangered or exterminated species.

The American Bison Society was established in 1905 with Hornaday as its president to preserve what was left of the bison herds. President Roosevelt encouraged preservation of the bison, and by 1908, the US Fish & Wildlife Service established the National Bison Range in Montana. Remnants of the herds were saved, and the warnings of the bison hunters proved not entirely in vain.

A *New York Times* editorial said Hornaday "deserves the gratitude and encouragement of the nation as the chief preserver from extinction of the

American bison" (1907, p. 8). Reaffirming the need for such efforts, Roosevelt's National Conservation Commission reported that game and fur-bearing animals were "largely exterminated" (National Conservation Commission, 1909). The movement to save the bison had gone from science to ethics to regulation in one short generation.

Protecting birds from "murder for millinery"

"To be fashionable, one must be extreme," a *Washington Post* writer declared in 1907. Women's hats should be "laden with feathers . . . in tremendous proportions" (p. 16). Feathers from flamingos, egrets, pelicans, pheasants, and crowned pigeons (goura) were especially favored.

And yet, if this meant beauty to some, it meant revolting cruelty to others. Starting in the 1870s, George Bird Grinnell and *Forest and Stream* magazine began to kindle a movement to stop the trade in feathers and the mass slaughter behind it.

Calls for restraint and bird protection laws flooded American newspapers and magazines by the 1890s and early 1900s. Three approaches made up the broad social movement:

- Ethical questions were raised by women's clubs and the Audubon societies that urged a boycott of fashionable feathered hats, often as an appeal to "higher womanhood" or avoidance of "murder for millinery" (Price, 1999);
- Scientific and educational groups, such as university ornithologists and the Smithsonian Institution, gave a view of the scope of the problem and the logic of reform;
- Advocates of government regulation, such as Roosevelt and Hornaday, provided a practical framework for reform.

An example of the ethical approach involved news coverage of a talk by Reverend O. P. Gifford, who described the egret plume in a woman's hat as a "brand of shame on thoughtless women" (*Washington Post*, 1897a, p. 7). Mark Twain's daughter, Jean L. Clemens, made "an earnest plea" to save the birds in 1906, writing that men "will shoot anything and everything that they can hit . . . Is there nothing to be done?" (*Baltimore Sun*, 1906, p. 10). And similarly, in 1899, then-governor of New York Theodore Roosevelt compared killing birds with vandalism.

> The destruction of the wild pigeon and the Carolina parquet has meant a loss as severe as if the Catskills or the Palisades were taken away. When I hear of the destruction of a species I feel just as if all the works of some great writer had perished.
>
> (*New York Times*, 1899, p. 7)

This ethical approach eventually led to the 1905 formation of the National Audubon Society from a variety of state level Audubon groups. The organization took the name of America's most famous artist and ornithologist, John James Audubon (1785–1871), whose *Birds of America*, printed between 1827 and 1838, is among the world's great treasures in print (Audubon, 1827).

The second approach, using an educational and scientific strategy, involved assessments of damage and logical arguments to the effect that birds could be beneficial because, for example, they eat insect pests. Edward Howe Forbush, an ornithologist who wrote *Useful Birds and Their Protection*, focused on their practical benefits and protection strategies. Institutions often added their weight to the educational arguments. The Smithsonian Institution supported this approach, as the *Washington Post* reported in 1897b. "It declares that civilized man is sweeping the wild birds off the face of the earth at such a rate that before long, hardly any species of feathered creatures will survive save those which are domesticated" (p. 17).

The third approach involved state and federal regulation to provide legal protection for threatened resources. US Representative John F. Lacey of Iowa stepped forward with the Lacey Bird and Game Act of 1900. The bill made transportation of bird skins and feathers a violation of federal law. Hat makers complained that the Lacey Act "has proven to be a serious blow" and threatened to put them out of business (*New York Times*, 1903, p. 3).

All three approaches worked together, but the ethical argument protected birds through peer pressure and effectively turned the tide. Women stopped wearing bird feathers in their hats and, by 1912, Paris fashion designers decreed that rare plumes would no longer adorn spring hats. According to the *New York Times*:

> Following the example of some American states, and for the first time in the history of Paris fashions, ostrich plumes and feathers of other beautiful and valuable birds are to be barred from the newest models of women's hats next season.
>
> (1912, p. C5)

The Audubon Society later suggested feathers from domestically raised birds as a substitute (*Washington Post*, 1940, p. 17).

The movement to protect birds led to human losses along the way. When two poachers killed Audubon game warden Guy Bradley on July 8, 1905, the news received wide coverage by the national press (*Baltimore Sun*, 1905). Poachers also killed game warden Columbus G. McLeod in Florida and Audubon Society employee Pressley Reeves in South Carolina in 1908. Twenty-five years later, retired *Miami Herald* editor Marjorie Stoneman Douglas wrote a short story, "Plumes," for the *Saturday Evening Post*, that described the emotions of Bradley's killer. "It was only weakness and sentimentality in him to flare up so at the idea of killing birds," she wrote (1930, p. 8).

Extinction of the passenger pigeon, 1914

Another major milestone in the news about extinctions was the death of Martha, the last passenger pigeon, in the Cincinnati Zoo in 1914. Her photo graces the first page of Hornaday's book, *Our Vanishing Wildlife*, which quotes ironically from an 1858 Ohio state report: "The passenger pigeon needs no protection. Wonderfully prolific . . . no ordinary destruction can lessen them, or be missed from the myriads that are yearly produced" (Hornaday, 1913).

As it turned out, the vast flocks of pigeons were decimated in a little less than 40 years, and again, the reaction was dismay at the slaughter (Greenberg, 2014). One Indiana farmer who had killed pigeons to protect his crops when he was young found that they had stopped coming as he grew older. He wrote in *Forest and Stream*: "Perhaps we were too prodigal as a people with our abundance, and a kind Providence took [the pigeons] away to teach us wisdom" (Hussey, 1914, p. 6).

Martha's death led to widespread introspection in the press.

> When we talk of extinct animals, we think of creatures like the mammoth or mastodon, which died out thousands upon thousands of years ago. We seldom realize that there are whole species which have vanished quite lately, and many which are at present just trembling in the balance.
>
> (*Washington Post*, 1915, p. MS 3)

Concerns about extinction led to the Migratory Bird Treaty of 1918 and the Bald Eagle Protection Act of 1940, yet lax enforcement meant that illegal traffic in wild bird plumage continued (*New York Times*, 1940, p. 23).

Conservation and the New Deal

As the dust settled after World War I, old heroes retired with honors and new issues arose to vex conservationists. In a 1925 White House ceremony, journalist and wildlife advocate George Bird Grinnell received the Roosevelt Memorial Association Medal for his work in wildlife conservation. Forester and Pennsylvania Governor Gifford Pinchot also received the award (*New York Times*, 1925, p. 23).

The new issues included a need to sort through the patchwork of state hunting regulations, and the election of President Franklin Delano Roosevelt in 1932 brought hope that the New Deal could coordinate conservation policies. Congress passed half a dozen conservation laws early in the New Deal, but conflicts over direct funding meant that they had to be set up with self-funding mechanisms. For example, improved habitats for ducks received funding through the federal "duck stamp" tax on duck hunting. In 1933, Roosevelt created a Migratory Bird Conservation Committee to help guide the programs, chaired by *Collier's* magazine editor Thomas H. Beck. The committee included wildlife management expert and ethicist Aldo Leopold, along with cartoonist Jay "Ding" Darling, who had set up a similar committee in Iowa (Cart, 1972, pp. 113–120).

Based at the *Des Moines Register*, "Ding" was a nationally syndicated cartoonist from 1900 to 1962 who earned two Pulitzer Prizes for cartooning. He also designed the first duck stamp. Ding's prominent conservation advocacy led Roosevelt to appoint him as head of the US Biological Survey in 1934, but he quit a year and a half later because Congress was "deaf to his appeal for funds." He also expressed dismay at the New Deal's still-disorganized conservation program:

> Conservation as a national principle has no substance or co-ordination. . . . Fourteen agencies in the Federal Government and forty-eight states . . . are like so many trains running on single-track roads, often in opposite directions and without any train dispatcher or block system. Collisions are frequent, wrecks are a daily occurrence, and the destruction is greater than the freight delivered at the specified destination.
>
> (Darling, 1935, p. 5)

After leaving the Biological Survey, Darling chaired a 1936 North American Wildlife Conference in Washington with the idea of unifying the forces he said were so confused. "In his ringing style of delivery, Darling drew a picture of polluted lakes and slaughter-house hunting grounds, the result of wild-life neglect and misuse," the *Washington Post* reported (1936, p. 15). The result was an organization initially called the General Wildlife Federation, with Darling as its first president. It became the National Wildlife Federation in 1938 and attracted millions of members in the following decades.

Another conservation-minded journalist in the mid-20th century was Edward Meeman, an editor with the Scripps-Howard newspaper chain in Tennessee and the chain's main conservation editor. He was known especially for his crusade on behalf of the Great Smoky Mountains National Park and support of the Tennessee Valley Authority's conservation programs. Other writers covering conservation in the interwar years included Richard Oulahan, whose 18-part series starting May 12, 1940, was about "nature and nature groups in Washington." The series began:

> Picture these United States without a tree. . . . The story of the destruction of America's great natural resources is a familiar one. On every hand are examples, mute testimony to the ruthlessness of human beings who despoiled and exploited the outdoor wealth of the continent. Acres of stumps where once were limitless forests; dust bowls born of fertile plains; birds animals and fish passing each year into extinction.
>
> (Oulahan, 1940, p. AM6)

World War II and postwar era conservation coverage

Conservation writing was a back-bench occupation during World War II, and although congressional hearings on wildlife conservation continued, they were rarely covered. One reporter observed wryly that far less hunting and fishing was going on in 1944, and that turned out to be a silver lining for the Fish and Wildlife Service conservation policy (McBride, 1944, p. S4). Occasional wildlife stories were framed in terms of the war effort in the 1940s. "During the pressure of war there is grave danger that conservation may lose ground; ground [that had been] gained through a hard struggle over a period of many years" (Oscher, 1943, p. L5).

The postwar years were a time for taking stock, and three books along those lines rose to prominence. *The Everglades: River of Grass* (1947) by former *Miami Herald* editor Marjory Stoneman Douglas described the history and wildlife of the Florida Everglades. Like Rachel Carson's *Silent Spring* 15 years later, the book became a classic of environmental literature.

Another, *Our Plundered Planet* by New York Zoological Society chief Fairfield Osborn (Osborn, 1948) was a Malthusian guide to post-war life. It was appreciated in later years for its early opposition to pesticides and as part of the birth of global ecology (Robertson, 2012).

William Vogt's 1948 *Road to Survival* is a summary of the world's diminished ecological status, describing US history as a "march of destruction." The *Washington Post* called him a "pious Malthusian" with a gift for sardonic epigrams (Lalley, 1948, p. 6). However, Vogt's comments on the decline of natural systems in Africa and Latin America were apropos, given the renewed US commitment to internationalism, and the establishment of the International Union for Conservation of Nature in 1948. The IUCN was not well-covered at first, except for a short *New York Times* story about the UNESCO conservation conference in Lake Success the next summer (1949, p. 31).

Later, as the problem of international conservation was acknowledged, John B. Oakes of the *New York Times* began addressing conservation as "A World Problem." He wrote: "The rapid disappearance of big game in America is, unfortunately, being paralleled in other continents right now" (1956, p. 101). As both a reporter and editorial page editor at the *Times*, Oakes brought conservation news to a higher level. He also was one of the few who made the transition to the more aggressive world of environmental reporting. In 1994, the Natural Resources Defense Council established an award for environmental journalism in his name.

The 1960s

The shift from "conservation writing" to "environmental journalism" occurred gradually over the 1960s, but one starting point was a 1961 speech by newly inaugurated President John F. Kennedy to the National Wildlife Federation. "It is our task," Kennedy said, "to protect those resources for future generations – to hand down undiminished the natural wealth and beauty which are ours today" (Casey, 1961, p. A2).

The charm of the Kennedy administration brought new stars to conservation causes, from aviation pioneer Charles Lindbergh (Whitman, 1969, p. 1) to Prince Phillip of Britain (Kihss, 1980, p. B3) to movie star Brigitte Bardot (Whittington, 1983, p. A1). All these and many more turned up in the media as advocates for endangered species during the late 20th century's environmental revolution.

The most celebrated was Rachel Carson, the author of *Silent Spring*, which was the most influential piece of environmental reporting of the era. A former writer with the US Fish and Wildlife Service, Carson was concerned about the unintended effects of pesticide misuse on birds. She struck a nerve even though the basic news about misuse of pesticides had already been covered (*Washington Post*, 1960, p. 1). At an August 1962 press conference, President Kennedy was asked whether government agencies should study the growing concern over DDT. He replied, "Yes, and I know that they already are, and I think particularly of course of Miss Carson's book, but they are examining the matter" (*New York Times*, 1962, p. 10). "This was a cleaving point," wrote Carson biographer William Souder, "the moment when the gentle, optimistic proposition called 'conservation' began its transformation into the bitterly divisive idea that would come to be known as 'environmentalism'" (2012, p. 5).

Like "Ding" Darling, many journalists who wrote in the gentle, optimistic voice of conservation began a transformation of their own. These included John B. Oakes, Gladwin Hill, and E. W. Kenworthy of the *New York Times*, syndicated columnist Drew Pearson, and *Washington Post* writer George Wilson (Neuzil & Kovarik, 1996). Pearson brought his nephew's concerns for wildlife into his columns as a nod to a younger generation's idea that saving animals was more important than politics (1964, p. B25).

As wildlife preservation became popular, it also became far more political. New controversies broke out about advocacy and environmental journalism. "Any media reporting of an environmental issue is advocacy, in the sense that media coverage draws attention to that issue," Philip Tichenor and Mark Neuzil wrote in *Nieman Reports* in 1994. "Airing a crisis that pits loggers against spotted owls lends the issue a measure of legitimacy" (Tichenor & Neuzil, 1996).

Ann Cottrell Free of the *Baltimore Sun* saw conservation coverage turning into a crisis. "Our giant of a nation is stirring," she wrote in 1969.

> It is in the early stages of a "Great Awakening." Not only are some of us wondering about the implications of [DDT in] eagle's eggs, many of us are realizing that our fouled air and water could soon come seeded with destruction for the human race.
>
> (p. WK6)

Oakes caught the new spirit of the environmental debate when he wrote in a 1979 column that the false and reckless charges over the economic impact of environmental lawsuits were evidence that "McCarthyism still lives" (p. A27). One issue he singled out was the 1978 Tellico Dam controversy, which "destroy[s] more in natural values than it can possibly create in artificial ones." The dam project in Tennessee had been stalled in part due to the presence of endangered snail darter fish. Congress approved construction of Tellico, overriding its own endangered species laws. However, Congress refused similar calls for exceptions to the law, especially for the New River in western Virginia and North Carolina. Kenworthy, another *Times* reporter whose main beat was the environment, began one article on the New River dam controversy by quoting US Supreme Court Justice Oliver Wendell Holmes: "A river is more than an amenity; it is a treasure" (1974).

Any big species on the verge of extinction is likely to be the focus of news reports, and that's especially true of charismatic megafauna like whales and dolphins. As early as 1910, a *New York Times* editorial endorsed the idea of whale conservation. "Now that the bison is saved from extinction," the paper said, scientists were calling for "the rescue of his warm-blooded cousin of the ocean plains" (*New York Times*, 1910, p. 8).

International studies about whale conservation began in 1924 with a League of Nations conservation committee, and League members signed a treaty pledging an effort to conserve whales (*New York Times*, 1932, p. XX3). A similar set of studies, followed by a similar international commitment, was reported in 1948 (*New York Times*, 1948, p. 28). When the International Whaling Commission acknowledged that blue whales were nearly extinct in 1965 (Bevona, 1965, p. A31), calls for a halt to whaling came not only from the commission and the IUCN, but also from conservation groups around the world (Hill, 1969, p. XX23). The US government banned importation of whale products in 1970 by listing whales as endangered species. Whaling was mostly banned internationally in 1986.

Little opposition to wildlife conservation is evident in the news media until the 1970s, when politically conservative reactions to environmental issues started popping up in the press. Concern for whales was nothing more than an "emotional squabble," wrote Norman Pearlstine of the *Wall Street Journal*. "Whale meat, in Japan, is poor man's beef," and banning whale hunting there would be like banning cattle ranching in America, he said (Pearlstine, 1974, p. 36).

Amid increasing acrimony, activists were determined to use the media to change the environmental debate. An upstart group from Vancouver, British Columbia, started using nonviolent tactics to put themselves between harpoons and the whales they hoped to save. "For the first time in the history of whaling, human beings had put their lives on the line for whales," wrote Charles Flowers (1975, p. 192), a book critic for the *New York Times*, who found attempts to save whales ironic historically, noting that in the 19th century, "American whalers were as bloodthirsty as any."

Flowers missed the point of the new activism. The idea, said Robert Hunter of Greenpeace, was to create a "mind bomb" – that is, an action that would create a dramatic new impression and replace an old cliché. In this case, that meant overturning the image of heroic whalers and replacing it with the image of heroic ecologists risking their lives to save the gentle giants of the sea. Greenpeace caught the world's attention and dramatically changed the political terrain for the media and the environmental movement (Kovarik, 2010). Partly as a result, public opinion research showed that environmentalism was "one of the most successful contemporary movements in the US and Western Europe" (Mertig & Dunlap, 1995, p. 145). Of course, whales were not the only charismatic species studied while under threat of extermination. Dian Fossey, a scientist and crusader for preservation of mountain gorillas in Africa, was famous for her 1983 book, *Gorillas in the Mist*. Her 1985 murder in Rwanda was widely covered in the world press (Battiata, 1986, p. C1).

News coverage expanded with more dramatic events occurring. "The late 1980s and early 1990s were salad days for the environmental beat," as Neuzil wrote in his book *The Environment and the Press* (2008, p. 196).

> With that growth came challenges and problems, some of which sounded familiar to those who had been in the field for decades . . . many stories still suffered from poor sourcing . . . some stories were superficial and under-reported; and dramatic events still took precedence over harder-to-understand environmental problems.

One antidote to superficial event-oriented news was the establishment of the Society of Environmental Journalists in 1990.

A harbinger of difficulties to come was Phil Shabecoff's forced retirement from the *New York Times*. "I resigned because I was taken off the environmental beat in 1990 because my coverage about things like climate change was considered alarmist," he said (Bravender, 2018).

Shabecoff's successor on the beat, Keith Schneider, was frequently criticized from both the conservative and liberal sides of environmental debates. In one particularly controversial article about dioxin, a toxic chemical blamed for human and animal deaths, Schneider wrote that exposure "is now considered by some experts to be no more risky than spending a week sunbathing" (1991, p. A1). The key word was "some" experts, and the article also included sharply critical points of view about dioxin from scientists with a different point of view, indicating that the controversy was unsettled. However, Schneider was widely criticized for including the "sunbathing" remark. *USA Today* environment editor Rae Tyson said: "I think Schneider messed up. It was not because he questioned a time-honored premise, but because he relied on some questionable science to prove his point" (Monks, 1994). In any event, the controversy benchmarked the increasingly complex field for environmental news coverage.

Meanwhile, just as environmental reporting dealt with more controversy and complexity, time and space for environmental stories began shrinking. In 2004, Sharon Friedman, a professor of science and environmental communication at

Lehigh University in the United States, warned that environmental news coverage was dropping quickly, especially in smaller newspapers. Still, the quality of environmental journalism was generally positive, she said (2004, p. 175).

In the first decades of the 21st century, environmental reporting faced a counter-movement against environmental science fueled by right-wing conspiracy theories. These included racist misinformation about DDT use (Milloy, 2007); misinformation about United Nations environmental treaties (Southern Poverty Law Center, 2014); and, especially, efforts to depict the science around climate change as fraudulent (Oreskes & Conway, 2010).

By the end of the second decade of the 21st century, the watchdog group Public Citizen warned that news about climate science in the right-of-center Fox News media was being deliberately crafted "to undermine climate science, cast climate advocates as hysterical and frame climate policy as dangerous and un-American" (Hartman, 2020). Worse, right-wing environmental advocates spread the racist belief that the root cause of environmental destruction is overpopulation by non-white people.

The media also suffered major setbacks to the capacity and credibility of environmental journalism in the early 21st century. CNN management, for example, dismissed its entire Science and Technology team in 2008 (Brainard, 2008). Another setback involved the dismantling of the *New York Times* environmental desk in 2013 (Brainard, 2013). In both cases, the explanation involved financial pressures. However, in both cases, sources close to the situation said the pressures were political, not financial. "They've made a horrible decision that ensures the deterioration of the *Times*'s environmental coverage at a time when debates about climate change, energy, natural resources, and sustainability have never been more important to public welfare," said Curtis Brainard of the *Columbia Journalism Review*.

Still, many news organizations did face significant financial problems. National Geographic Society's *The Green Guide* ceased publication in 2008; *E Magazine* stopped print production in 2013; *Pacific Standard* closed in 2019; and an overall decline in coverage reflected the financially distressed media environment of the early 21st century.

By the 2020s there was reason for hope. Support for the environment in public opinion polls held steady at over 70 percent (Gallup, 2021). The *New York Times* had renewed its commitment to environmental coverage by establishing a climate desk in 2017, and, during the 2020 election, a collaborative effort by 400 publications covered the impacts of climate change worldwide (Tracy, 2019, p. B1; Hertsgaard & Pope, 2020). The idea was to "tell the climate story so people can get it," a *Columbia Journalism Review* article said (Hertsgaard & Pope, 2019). Mark Hertsgaard, the collaborative's editor, said:

> Despite the "climate silence" that had long prevailed in much of the media, especially in the US, we sensed that there *was* a critical mass of journalists and news outlets that recognized the need to do better. If we could highlight this critical mass, we thought, we could grow it.

Conclusion

This historical portrait of news about endangered species shows that media coverage changed through several historical periods. During the Progressive movement around the turn of the 20th century, the press responded to the facts of extinction by covering scientific discussions, ethical movements, and debates over public policy responses. As a part of the New Deal of the 1930s, several major figures in the press joined the Roosevelt Administration to help guide public policy. A consensus about conservation meant that little public controversy took place, and writers focused on the decline in resources, for example, at the start of *Washington Post* writer Richard Eulachan's 1940 series: "Picture these United States without a tree."

Coverage changed again from the post-war years to the 21st century, as the underlying focus shifted from declining natural resources to the link between wildlife and human survival. For example, the *Baltimore Sun's* Ann Cottrell Free warned in 1969 that "fouled air and water could soon come seeded with destruction for the human race." More than fifty years later, news organizations carried similar warnings from scientists and international officials about the link between biodiversity and human existence, for example, in this March 2021 op-ed carried by CNN: "The scientific evidence calls for a complete change – a total overhaul of our relationship with nature and biodiversity. This is not an option, but a necessity for our survival" (Azoulay et al., 2021).

References

Audubon, J. J. (1827–1838). *Birds of America*. Edinburgh: Havell.

Azoulay, A., Lee, H., Hernández Salgar, A. M., Thomson, P., Goodall, J., Princess Sumaya of Jordan, Ibrahim, H. O., & Hulot, N. (2021, March 24). Our survival depends on treating nature with more respect. CNN. Retrieved from https://www.cnn.com/2021/03/24/opinions/biodiversity-nature-survival-azoulay-lee-salgar/index.html

Baltimore Sun. (1905, August 6). His life to save birds.

Baltimore Sun. (1906, December 26). Wants birds saved.

Battiata, M. (1986, January 25). Murder and the private war: Dian Fossey. *Washington Post*.

Bevona, D. (1965, May 2). Blue whale nearly extinct. *Washington Post*.

Brainard, C. (2008, December 4). CNN cuts entire science, tech team. *Columbia Journalism Review*. Retrieved from https://archives.cjr.org/the_observatory/cnn_cuts_entire_science_tech_t.php

Brainard, C. (2013, March 1). *New York Times* cancels Green blog. No explanation from editors following surprise announcement. *Columbia Journalism Review*. Retrieved from https://archives.cjr.org/the_observatory/new_york_times_cancels_green_e.php

Bravender, R. (2018, December 14). Pioneering environment writer wants an apology. *E&E News*. Retrieved from www.eenews.net/special_reports/offtopic/stories/1060109689

Briggs, H. (2020, June 2). Extinction crisis poses existential threat to civilization. BBC. Retrieved from https://www.bbc.com/news/science-environment-52881831

Bruner, F. (1935, March 23). Richberg taking NRA helm. *Washington Post*.

Carr, L. (1874). *Judith Gwynne*. London: H. S. King & Co.

Carroll, L. (1865). *Alice's adventures in Wonderland*. London: Palgrave Macmillan.

Cart, T. W. (1972). New deal for wildlife. *Pacific Northwest Quarterly, 63*(3), 113–120.
Caryn, J. (1989, July 16). Are feminist heroines an endangered species? *New York Times.*
Casey, P. (1961, March 4). President pleads for wildlife. *Washington Post.*
Darling, J. N. (1935, October). Desert makers. *Country Gentleman, 105.*
De la Beche, H. T. (1831). *A geological manual.* London: Treuttel and Würtz.
Dixon, G. (1962, July 4). The vanishing American male. *Washington Post.*
Douglass, M. S. (1930, June 15). Plumes. *Saturday Evening Post.*
Douglass, M. S. (1947). *The Everglades: River of grass.* New York: Rinehart & Co.
Flowers, C. (1975, August 24). Between the harpoon and the whale. *New York Times.*
Fossey, D. (1983). *Gorillas in the mist.* Boston, MA: Houghton Mifflin.
Free, A. C. (1969, April 13). The great awakening: City hall, Congress and the White House. Are aware of our battle for survival. *Baltimore Sun.*
Friedman, S. (2004). And the beat goes on: The third decade of environmental reporting. In S. Senecah (Ed.), *Environmental communication yearbook* (Vol. 1). Mawah, NJ: Lawrence Erlbaum.
Fuller, E. (2002). *Dodo – from extinction to icon.* London: Harper Collins.
Gallup Inc. (2021). Environment. Retrieved from https://news.gallup.com/poll/1615/environment.aspx
Glacken, C. (1976). *Traces on the Rhodian shore.* Berkeley, CA: University of California Press.
Greeley, H. (1860). *An overland journey from New York to San Francisco, in the summer of 1859.* New York: C.M. Saxton Barker & Co.
Greenberg, J. (2014). *A feathered river across the sky.* New York: Bloomsbury.
Hartmann, B. (2020, Spring). The ecofascists. *Columbia Journalism Review.* Retrieved from www.cjr.org/special_report/ecofascists.php
Hertsgaard, M., & Pope, K. (2019, July 26). Can we tell the story so people get it? *Columbia Journalism Review.* Retrieved from www.cjr.org/covering_climate_now/covering-climate-partnerships.php
Hertsgaard, M., & Pope, K. (2020, September 23). The media's climate coverage is improving but time is very short. *Columbia Journalism Review.* Retrieved from www.cjr.org/covering_climate_now/the-medias-climate-coverage-is-improving-but-time-is-very-short.php
Hill, G. (1969, January 19). Cetacean society has whale of a job. *New York Times.*
Holder, C. F. (1899, September 9). The crime of a century. *Scientific American.*
Hornaday, W. T. (1913). *Our vanishing wildlife.* New York: Charles Scribner's Sons.
Humanitas. (1828, July 3). To the editor of the "Times." *Times of London.*
Hussey, T. (1914, March 21). Days of passenger pigeons. *Washington Post.*
Jefferson, T. (1787, April). Comparative view of the animals of America and those of Europe. *Columbian Magazine.*
Jewett, S. (1886). *A white heron and other stories.* New York: Houghton Mifflin.
Kant, I. (1784 [1997]). *Moral philosophy: Collin's lecture notes* (P. Heath & J. B. Schneewind, Eds. & Trans., pp. 37–222). Cambridge, UK: Cambridge University Press.
Kenworthy, E. W. (1974, December 8). The nation in summary. *New York Times.*
Kihss, P. (1980, October 2). Plea for wildlife made in speech by Prince Philip. *New York Times.*
Kovarik, B. (2010). Greenpeace. In S. H. Priest (Ed.), *Encyclopedia of science and technology communication.* Los Angeles: Sage.
Lalley, J. M. (1948, August 16). The books that last. *Washington Post.*
Leopold, A. (1949). *A Sand County almanac and here and there.* New York: Oxford University Press.

McBride, T. M. (1944, March 12). War may be aiding fish, wildlife service's conservation policy. *Washington Post.*

Melville, H. (1851). *Moby Dick or the whale.* New York: Harper & Brothers.

Mertig, A., & Dunlap, R. (1995). Public approval of environmental protection. *International Journal of Public Opinion Research*, 7(2).

Milloy, S. (2007, June 9). The malaria clock: A green eco-imperialist legacy of death. *Junk-science.com.*

Monks, V. (1994, August). The truth about dioxin. National Wildlife Federation. Retrieved from www.nwf.org/Magazines/National-Wildlife/1994/The-Truth-About-Dioxin

National Conservation Commission. (1909). Report of the National Conservation Commission. 60th Congress, 2nd Session, Senate Document No. 676. Washington, DC: US Government Printing Office.

Neuzil, M. (2008). *The environment and the press: From adventure writing to advocacy.* Evanston, IL: Northwestern University Press.

Neuzil, M., & Kovarik, W. (1996). *Mass media and environmental conflict, America's green crusades.* Thousand Oaks, CA: Sage.

New York Times. (1869, September 21). Amusements, theatrical, Miss Bateman as Lear.

New York Times. (1874, January 2). Some new stories about animals.

New York Times. (1875, August 2). Extinct characters. Reprinted from the *Richmond Whig.*

New York Times. (1882, March 19). Entire races extinct.

New York Times. (1884a, May 4). Newsy foreign gossip. Reprinted from the *Edinburgh Scotsman.*

New York Times. (1884b, September 18). Science to be surprised.

New York Times. (1888, January 22). Going to join the dodo.

New York Times. (1896, July 26). Killing of the bison.

New York Times. (1899, March 24). To protect the birds.

New York Times. (1903, February 16). Bird dealers protest.

New York Times. (1907, November 3). Bison preserves.

New York Times. (1910, November 1). Conservation of whales.

New York Times. (1912, January 28). Rare plumes won't adorn spring hats.

New York Times. (1924, November 16). Organ grinders and monkeys.

New York Times. (1925, May 16). Coolidge presents Roosevelt medals.

New York Times. (1932, April 24). Nations signing whale treaty pledge conservation efforts.

New York Times. (1940, October 16). Illegal traffic in plumage scored.

New York Times. (1948, November 27). World whale pact is ratified by US.

New York Times. (1949, August 24). Nature umpire urged on animal, plant life.

New York Times. (1962, August 30). Transcript of the president's news conference.

Niles Register. (1817, March 22). Sagacity of a dog, pp. 12, 60.

Niles Register. (1836, October 1). Value of a dog, pp. 51, 80.

Niles Register. (1845, September 20). Capt. Frémont's report, pp. 45, 69.

Oakes, J. B. (1956, July 18). Conservation: A world problem. *New York Times.*

Oakes, J. B. (1979, December 28). Saving the web of life. *New York Times.*

O'Malley, P. (1980, July 10). Girls keeping fast-pitch alive. *Washington Post.*

Oreskes, N.,& Conway, E. M. (2010). *Merchants of doubt: How a handful of scientists obscured the truth on issues from tobacco smoke to global warming.* New York: Bloomsbury.

Osborn, F. (1948). *Our plundered planet.* New York: Little Brown & Co.

Oscher, P. H. (1943, May 9). The squandered wealth of a nation. *Washington Post.*

Oulahan, R. (1940, May 12). Wildlife Federation fights to save nation's heritage. *Washington Post.*

Owen, R. (1841, July). *Report on British fossil reptiles. Part II*. Report of the eleventh meeting of the British Association for the Advancement of Science.

Pall Mall Gazette. (1886, February 11). The old dodo at Scotland Yard.

Pearlstine, N. (1974, June 26). Wailing over whaling: Angry Japanese see effort to stop killing. *Wall Street Journal*.

Pearson, D. (1964, July 20). Man to blame for extinct species. *Washington Post*.

Price, J. (1999). *Flight maps: Adventures with nature in modern America*. New York: Basic Books.

Robertson, T. (2012, April). Total war and the total environment: Fairfield Osborn, William Vogt, and the birth of global ecology. *Environmental History, 17*, 336–364.

Roosevelt, T. (1893). *Hunting the grisly and other sketches*. New York: G. P. Putnam's Sons.

Rosenberg, G. D. (2009). *The revolution in geology from the renaissance to the enlightenment*. McLean, VA: Geological Society of America.

Rupke, N. (1987). *Vivisection in historical perspective*. London: Croom Helm.

Schneider, K. (1991, August 15). US Backing away from saying dioxin is a deadly peril. *New York Times*.

Secord, J. (2000). *Victorian sensation: The extraordinary publication, reception and secret authorship of vestiges of the natural history of creation*. Chicago, IL: University of Chicago Press.

Sewell, A. (1877). *Black Beauty, his grooms and companions, the autobiography of a horse*. London: Jarrold & Sons.

Shepard, R. (1967, December 22). The movie newsreel flickers its last and dies. *New York Times*.

Souder, W. (2012). *On a farther shore; the life and legacy of Rachel Carson*. New York: Crown.

Southern Poverty Law Center. (2014). Agenda 21: The UN, sustainability and the right wing conspiracy theory. Retrieved from www.splcenter.org/20140331/agenda-21-un-sustainability-and-right-wing-conspiracy-theory

Starbuck, A. (1856). The story of the whale. *Harper's, 12*.

Strickland, H. E., & Melville, A. G. (1848). *The dodo and its kindred or, the history, affinities, and osteology of the dodo, solitaire, and other extinct birds of the islands Mauritius, Rodriguez and Bourbon*. London: Reeve, Benham, and Reeve.

Tichenor, P., & Neuzil, M. (1996, Winter). Advocacy, the uncomfortable issue. *Nieman Reports, 41*.

Times of London. (1824, June 17). Society for the prevention of cruelty to animals.

Times of London. (1832, August 29). Cuvier.

Times of London. (1835, January 24). Cuvier's animal kingdom.

Times of London. (1848, September 14). London is a city.

Times of London. (1853, May 25). There are two things in antiquity.

Times of London. (1875, June 9). The late Comte de Remusat.

Tracy, M. (2019, July 8). Newsrooms face a changing climate. *New York Times*.

Turnbull, A. (Ed.). (1964). *The letters of F. Scott Fitzgerald* (1st ed.). London: Bodley Head.

Vogt, W. (1948). *Road to survival*. New York: William Sloane.

Wallace, A. R. (1880). *Island life: Or, the phenomena and causes of insular faunas and floras, including a revision and attempted solution of the problem of geological climates*. London: Richard Clay and Sons.

Washington Post. (1878, February 18). Personal.

Washington Post. (1888, August 24). Buffalo and elk will not become extinct.

Washington Post. (1889, July 21). The talking machine.

Washington Post. (1890, January. 4). A war of extermination.

Washington Post. (1897a, February 14). Taking final flight.

Washington Post. (1897b, April 12). Pleads for the birds.

Washington Post. (1902, November 15). Tied up a bear but failed to shoot it. Reprinted from the *Baltimore Sun*.

Washington Post. (1907, September 24). Fall opening at Kann's.

Washington Post. (1915, April 4). Animals man has wiped out.

Washington Post. (1936, February 4). Darling talks at wild life session.

Washington Post. (1940, December 13). Bird massacre for millinery opposed.

Washington Post. (1951, October 22). Missouri mule on the wane.

Washington Post. (1960, December 27). Pesticides blamed: Census finds bald eagle vanishing from capital.

Whitman, A. (1969, June 23). Lindbergh traveling widely as conservationist. *Washington Post*.

Whittington, L. (1983, February 24). Europe's threat halts Canadian seal pup hunt. *Washington Post*.

Wilson, A. (1950). *Such darling dodos*. London: Secker & Warburg.

3 The extinction crisis

Why words matter

Michael A. Patten and Brenda D. Smith

If we could transport ourselves to the not-so-distant past – a mere four or five centuries would suffice – many of us might be surprised to discover that *rhetoric* was not always an undesirable commodity. We think of rhetoric now in negative terms, whether accusing a disliked politician who employs rhetoric of being disingenuous or duplicitous or using it as a derogatory word in an argument, as if someone who engaged in rhetorical speech lacked key moral qualities. Yet in our shared, not-so-distant past, rhetoric was an esteemed art, a demonstration that one could combine emotion, logic, and credibility to persuade, inform, or influence others. Plato emphasized pursuit of truth, the dialectic imperative, as the goal of rhetoric. It is a noble goal indeed.

Despite its slip into slur, its near-universal association with unsavory political trickery or obfuscation, rhetoric remains an important faculty. It is especially important in fields that demand a certain level of public engagement to thrive, such as conservation biology, which must have broad support to have any efficacy. If a large enough proportion of the general public is heedless of the welfare of other species or the future of Earth's wild lands and ecosystems, then we can expect little beyond minor, local victories in the ongoing battle to conserve biodiversity.

It is then obvious that rhetoric, the art of persuasion, plays a key role in conservation biology. Even if we dodge the label, those of us who wish to protect our natural heritage use Aristotelian rhetoric when we strive to win over hearts and minds.

The question is what happens when we ourselves are guilty of lending credence to the baggage the term now carries? Are there times when our rhetoric is so ill-conceived or poorly considered that we accomplish nothing, apart from affirming negative connotations? Put another way, have we maintained Aristotle's balance (Figure 3.1) among logic, emotion, and credibility or do we skew heavily towards one pole?

The authors suggest that the answer to these questions is a qualified "yes," that we too often emphasize emotion over logic and credibility when we attempt to convert the masses to the conservation gospel. A telling aspect of appeals to emotion is how we use words. The pedant who is overly concerned with rules or trivial details in each of us wants words to have meaning,

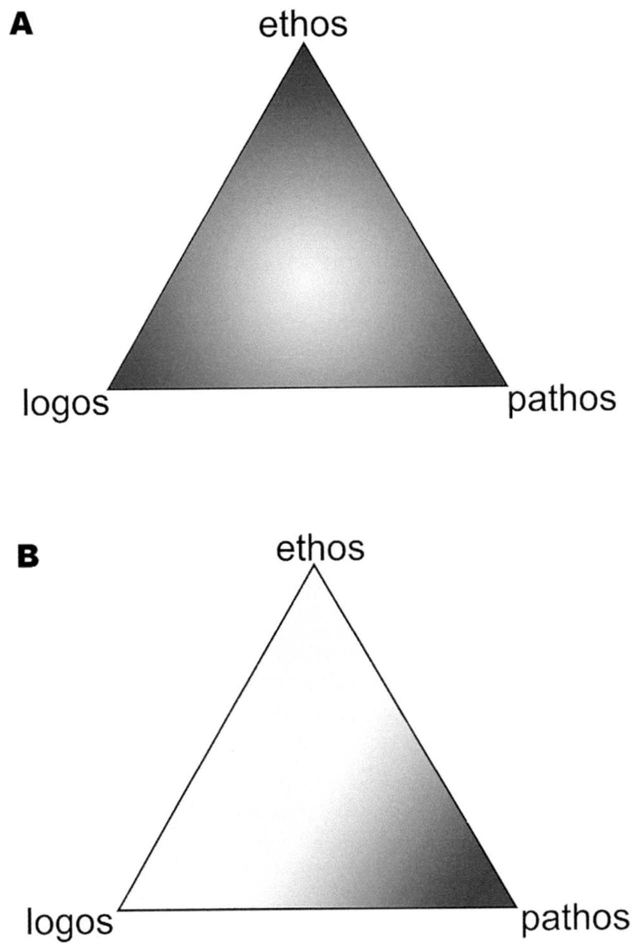

Figure 3.1 Aristotle's rhetorical triangle. A) In his conception, the three poles – logos (logic; e.g., supporting data), ethos (credibility; e.g., knowing the material), and pathos (emotion; e.g., appeal to morality) – received equal weight; no pole was more important than the others. B) Modern use of rhetoric, perhaps influenced disproportionately by the advent of advertising (Harris & Sanborn, 2014), has apparently tilted heavily towards the pathos pole.

Source: Image courtesy of Michael A. Patten and Brenda D. Smith.

yet we each appreciate that words have power to persuade. In rhetoric, words often become signifiers of a larger concept or a broader appeal. Such words may be used repetitively as "commonplaces" (Ross, 2013), in which case the general meaning is assumed to be shared by members of the audience or community.

Statement of the problem

A prevalent commonplace in conservation rhetoric is *extinction*, a word whose meaning is at once intuitive and slippery, the latter because it often is modified as if it were something other than an absolute (Smith-Patten et al., 2015). The International Union for Conservation of Nature (IUCN) defines *extinction* clearly enough: It applies only to those species for which there is "no reasonable doubt that the last individual has died." A child knows that the great dinosaurs are extinct and uses the term in its proper, IUCN sense. Then how has the word become so freighted with sly and subtle shading in much of the conservation biology literature? Why has this absolute been modified so often – "locally extinct" or "regionally extinct" or "truly extinct," to name a few real-life examples – as to become the "very unique" or "more perfect" of the biological sciences?

The fault may, in part, lie with how press releases – a form of advertising – are shaped by public relations offices and repackaged by media outlets (Ladle, Jepson, Araújo, & Whittaker, 2004; Smith-Patten et al., 2015), but it seems unfair to lay blame at the feet of players outside of the researchers' realm. Realistically, media forces routinely promulgate misleading terminology but are seldom the source of it. Consider this example of shape-shifting that Smith-Patten et al. (2015) outlined: A research paper whose title did not mention extinction, "Pervasive defaunation of forest remnants in a tropical biodiversity hotspot" (Canale, Peres, Guidorizzi, Ferreira Gatto, & Kierulff, 2012), spawned an edgier press release titled "Widespread local extinction in tropical forest 'remnants,'" with the authors asserting "that remaining habitat fragments had been *virtually emptied of their forest wildlife*" [emphasis added]. The media dropped the modifier and ran with the "virtually emptied" storyline, leading to headlines such as *Brazil's Rainforest Suffering Widespread Extinction* and *In Fragmented Brazilian Forest, Few Species Survive* (Smith-Patten et al., 2015). This example, one of many, illustrates how quickly science can morph into misinformation. It began with an impartial study from whence sprang a sensational media feed replete with statements that likely would not pass peer review. From there, the story morphed quickly into alarming headlines as subtleties were elided in favor of rhetorical punch.

Hence the fault also lies with conservation biologists themselves. It may feel unfair to blame a conservation biologist who employs rhetoric, even without overt knowledge of doing so, to draw attention to his or her important work, but it remains that words matter, that terms must stand for something. If the term *locally extinct* is used in a scientific article, a conservation biologist may protest innocence and may assert that "readers will know my meaning." That is true, so far as other scientists versed in biological theory go. The problem is that *locally extinct*, a modified absolute, does not translate well to common parlance. It is too easy for a reporter, just as innocently and with an eye towards deadlines and rhetorical demands, to excise what is seen as a needless modifier. By doing so, the reporter emphasizes the money word, *extinct*. Where does fault lie then? Should the reporter have known better? If so, how? Or should the researcher have made the modified term's meaning clear by stating unequivocally that *locally extinct* does not mean "no reasonable doubt that the last individual has died?" This argument is not a semantic one: The modifier changes everything.

The likely culprits are colleagues who succumb to their own rhetorical demands and thereby insist on using *extinction*, albeit in a modified form, even though a specific word is available: *extirpation*, meaning an organism or species ceases to exist in a particular area. Over the decades, *extirpated* and *extirpation* have been supplanted by *extinct* and *extinction*, modified or not (Smith-Patten et al., 2015). Usage frequency fluctuates greatly for many words – it is a natural part of language evolution and associated cultural change (i.e., culturomics; Ladle et al., 2016). Nowadays, the word *extirpate* is not commonly known or used among members of the general public and the press, while the word *extinct* is known, used widely, and grabs attention. It more deeply satisfies needs of the pathos pole of Aristotle's rhetorical triangle. After all, "[s]cientists, environmentalists, and politicians have repeatedly used extinction rhetoric as a core justification for a global conservation agenda that seeks to influence a wide range of human activities" (Ladle & Jepson, 2010, p. 96).

Lamentably, "extirpation rhetoric" lacks a similar appeal to pathos. It may be futile to argue that a more accurate word is eschewed for a less accurate word. The cachet is not the same, and there lies the problem. Any rhetoric-based effort to generate support for the biodiversity cause necessarily will fall back on commonplaces that strike a nerve, any nerve.

Social implications of overuse and misuse of *extinction*

In a world where so many pressing issues scream for attention, those who scream the loudest get the most attention. For several decades, the assumption was that screaming "*extinction!*" in earnest was akin to screaming "*fire!*" in earnest – people were expected to react immediately to an obvious danger. Instead, the public has grown weary, confused, and complacent. "*Extinction!*" has been screamed so often that the thought of it is too often relegated to the background; it has been reduced to white noise.

As a term, *extinction* has been misused so often that it no longer denotes its original meaning, at least not fully. Any person who retains a notion of the word's meaning can be forgiven for her or his confusion when, for example, an online search yields many results that conflate *extirpation* and *extinction*, yet use only the latter term. One such result asked and answered "Is Extinction forever? Well not in Botswana." The article explained that "every last" rhino was lost in the 1980s, after which it was re-introduced but "went extinct in Botswana for a second time" in the 1990s (www.wildlifefilms.co/rhino-rescue). In this vein, it is plausible to posit that someone could read an alarming article that a bird or mammal has been declared "extinct," only to find an article months later that trumpets how the species is set to benefit from millions of tax dollars for its preservation. Perhaps worse, the general public may sense that the system is rigged and may believe that the word *extinct* is used not just to draw attention to biodiversity's plight but to attract grant money and generate publicity. How can such confusion or such cynicism not foster a mistrust of science?

Yet a peculiarity of human behavior, that of at once mistrusting something while putting full faith in it, comes to play in the ethical dilemmas of conservation. Humans both fear and revere science and technology. Back to the late 18th century, when Thomas Malthus posited that the human population would overwhelm its capacity to feed itself, people have simultaneously feared science (e.g., innovative crop production such as genetic modification) and trusted that science will somehow save them.

Our duality is evident, too, with issues such as climate change and the extinction crisis. Many people recoil from ethical dilemmas where prospective solutions of these vexing challenges arise. One who wishes to reduce her or his environmental impact is told, for example, to lessen reliance on fossil fuels and to limit reproduction. However, heeding such dictums engenders a great deal of personal sacrifice. Those who blanch at the sacrifice may reject the science or rationalize their behavior via the false notion that individual actions have too little effect on global problems. More insidiously, they do nothing as they await a scientific breakthrough to save the day. In essence, they hope someone else will solve the world's problems, likely missing the irony that they mistrust science while simultaneously banking on science to save the day.

Some believe extinction will have a "biotechnological savior" in the guise of the emerging, and controversial, study of "de-extinction" (Jørgensen, 2013; Cohen, 2014; Richmond, Sinding, & Gilbert, 2016; Seddon, 2017). The notion that an extinct species may be "resurrected" is a conservation biologist's nightmare. It is not because the conservation biologist wants any species to be extinct but because the entire concept is likely to sap what little public support remains to address the biodiversity crisis. In effect, it is a nightmare of reliance upon technology to "save" a species after its extinction instead of preventing its extinction in the first place. Reliance on "de-extinction" presents us with a hypothetical world full of genetically engineered, reanimated species that would be arrogantly plugged into ecosystems where we think their ecological functions lay, although we are usually ignorant of those traits. De-extinction is not a savior but rather the indolent person's refuge. It would be an amazing technological feat, but it would grant a free pass to the very behavior that leads to extinction: not thinking beyond one's short-term actions and so gaining the privilege to waive away long-term consequences.

If we build ourselves an ethical escape hatch, then conservation biology becomes a science no longer aimed at providing solutions to the looming loss of species but instead becomes the mundane task of documenting species' first demise. Continuing to allow the finality of extinction to get lost amid conflated rhetoric and false notions of a world easier to recreate than to preserve would further the general complacency towards the extinction crisis and, in so doing, lose the most important rhetorical battle of all.

Acknowledgments

This chapter draws on research previously published in Smith-Patten, B.D., Bridge, E.S., Crawford, P.H.C., Hough, D.J., Kelly, J.F., & Patten, M.A. (2015). Is extinction forever? *Public Understanding of Science*, *24*(4), 481–495.

References

Canale, G. R., Peres, C. A., Guidorizzi, C. E., Ferreira Gatto, C. A., & Kierulff, M. C. M. (2012). Pervasive defaunation of forest remnants in a tropical biodiversity hotspot. *PLoS One*, 7(8), e4167. doi:10.1371/journal.pone.0041671

Cohen, S. (2014). The ethics of de-extinction. *Nanoethics*, *8*, 165–178. doi:10.1007/s11569-014-0201-2

Harris, R. J., & Sanborn, F. W. (2014). *A cognitive psychology of mass communication* (6th ed.). New York: Routledge.

Jørgensen, D. (2013). Reintroduction and de-extinction. *BioScience*, *63*, 719–720. doi:10.1525/bio.2013.63.9.6

Ladle, R. J., Correia, R. A., Do, Y., Joo, G. J., Malhado, A. C. M., Proulx, R., . . . Jepson, P. (2016). Conservation culturomics. *Frontiers in Ecology and the Environment*, *14*, 269–275. doi:10.1002/fee.1260

Ladle, R. J., & Jepson, P. (2010). Origins, uses, and transformation of extinction rhetoric. *Environment and Society*, *1*, 96–115. doi:10.3167/ares.2010.010105

Ladle, R. J., Jepson, P., Araújo, M. B., & Whittaker, R. J. (2004). Dangers of crying wolf over risk of extinctions. *Nature*, *428*, 799. doi:10.1038/428799b

Richmond, D. J., Sinding, M. H., & Gilbert, M. T. (2016). The potential and pitfalls of de-extinction. *Zoologica Scripta*, *45*, 22–36. doi:10.1111/zsc.12212

Ross, D. G. (2013). Common topics and commonplaces of environmental rhetoric. *Written Communication*, *30*, 91–131. doi:10.1177/0741088312465376

Seddon, P. J. (2017). The ecology of de-extinction. *Functional Ecology*, *31*, 992–995. doi:10.1111/1365-2435.12856

Smith-Patten, B. D., Bridge, E. S., Crawford, P. H. C., Hough, D. J., Kelly, J. F., & Patten, M. A. (2015). Is extinction forever? *Public Understanding of Science*, *24*, 481–495. doi:10.1177/0963662515571489

4 The wolves of fate

Media coverage of the Isle Royale "genetic rescue"

Alexander Killion, Mark Neuzil and Eric Freedman

Isle Royale, which sits in Lake Superior and belongs to the United States through the state of Michigan, is the largest in a group of more than 400 islands that make up Isle Royale National Park (IRNP). Its main draws are its wilderness and wildlife. Moose (*Alces alces*) and gray wolves (*Canis lupus*) have been the star attraction, yet their populations have waxed and waned over time, changing the surrounding ecosystem. By 2016, only two highly inbred wolves remained there. In a controversial move, the National Park Service (NPS) moved four gray wolves from the mainland onto the island in 2018. This was the first of several planned releases to artificially re-balance the wolf–moose populations, and it is only the second time that the NPS intervened in a designated wilderness area to manipulate a predator–prey relationship. The agency announced plans to relocate 20–30 wolves to Isle Royale in the first three years of the project and spend about $2 million over 20 years to maintain the wolf population (National Park Service [NPS], 2018). Supporters call this plan a "genetic rescue," but skeptics say nature should be allowed to take its course.

More people visit Yellowstone National Park on a single summer day than Isle Royale in an entire season, which runs from May to October (NPS, 2019a). Despite its isolation – or perhaps because of it – wolves and moose are the charismatic megafauna (i.e., large animals with symbolic value or widespread appeal) whose fate on Isle Royale has attracted widespread public interest and media attention for decades. Both species have cultural meanings that extend beyond their predator–prey relationship, yet their populations have fluctuated between peaks and crashes. As a result, officials have moved gray wolves in the Great Lakes region (Minnesota, Wisconsin, and Michigan) on and off the US Endangered Species List eight times since 2000.

From the time the NPS released its Record of Decision on June 7, 2018, announcing the wolf management plan for Isle Royale, the agency has provided detailed information to the media about the project while prohibiting a physical media presence on the island during the actual releases. The remoteness of these wolf translocations (i.e., the capture, transport, and release of an animal from one location to another) further inhibited media access to the island. All reporting had to be done by phone, email, through mainland press conferences and releases, and through social media sites. Without reporters' boots on the ground during the initialization of the management plan, this study had two research questions: 1) What media strategies and techniques did the NPS and its partners use to keep the press informed? 2) How did access restrictions affect the way news outlets covered the translocations?

Isle Royale: moose and wolves

The 205-square-mile (53,095 hectare) island is a nearly self-contained ecosystem where wolves and moose have existed in countable numbers, waxing and waning over the years (Figure 4.1). The island is in the middle of the largest freshwater lake on the continent and accessible by boat, seaplane, or ski plane during mild weather. There are no roads or motors, and no one stays over the winter except park employees and scientists.

Moose probably first migrated there in 1912–1913 during an unusually cold winter that solidified an ice bridge to Canada. Later, DNA evidence indicated that perhaps a Minnesota hunting club had stocked moose in the early 1900s. As wolves had not arrived by then, the moose population grew unchecked by predators. Adolph Murie, the first American scientist to study wolves in their natural habitat, estimated there were more than 1,000 moose on Isle Royale in 1929 (Grooms, 2002). As a result, vegetation favored by moose was

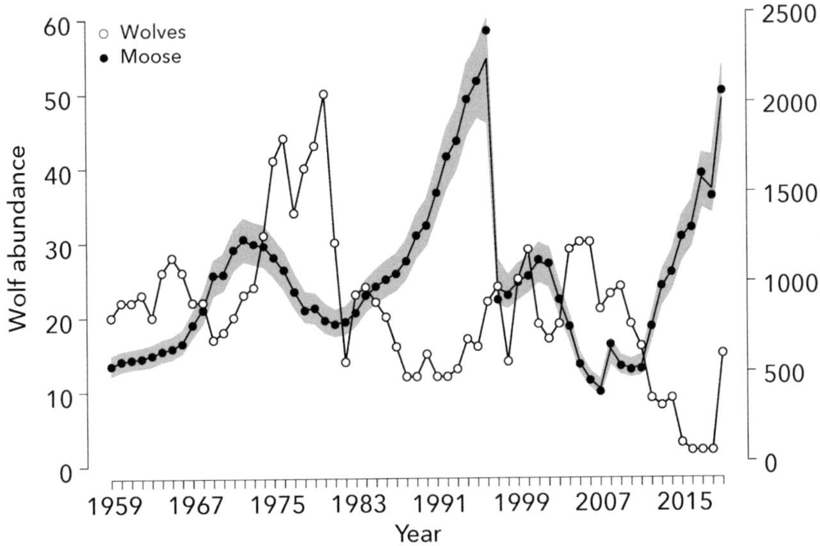

Figure 4.1 Wolf and moose population estimates in Isle Royale National Park from 1959 to 2019. Gray-shaded areas are 95% confidence intervals of the moose population. Wolf and moose values are represented on different scales to show how relative changes in the wolf population cause a change in the opposite direction for the moose population.

Source: Graph courtesy of Hoy, Peterson, & Vucetich, 2019, published with permission.

over-browsed. In the 1930s, a Michigan Department of Conservation photographer filmed the moose and spread their images around the world, thrusting them into charismatic megafauna status.

Concerned with overpopulation, Murie became the first of many scientists to suggest introducing a predator to the ecosystem, but only after rejecting hunting, either by sportsmen or the state. However, wolves were not translocated – perhaps due to the Great Depression and World War II – but instead found Isle Royale on their own. By some accounts, the first pair walked there over the ice in 1949 from Ontario (Line, 1996). In 1956, a census estimated at least 22 wolves lived on the island; by 1957, it was thought wolves may have pushed out the last coyote (Cole, 1957). In 1958, a close study of the predator–prey relationship began that continues today (Mech & Breining, 2020).

The mid-1960s marked a low point for the larger shore-bound Great Lakes wolf population. State bounties on wolves were ending but this was almost too late to assist in the recovery of the population. With only a few hundred wolves left in the continental United States, arguments that wolves had a detrimental impact on game animals or livestock were less compelling. Because of their small numbers or changing public attitudes, wolves began to creep back into public favor.

Michigan protected the wolf under its state endangered species law in 1965; in 1973, it was included in the first federal Endangered Species Act (ESA). Animals listed in the ESA were required to have a recovery plan, but the Isle Royale wolf population was excluded from the plan because of its isolation (US Fish and Wildlife Service, 2011). Groups such as the Audubon Society, National Wildlife Federation, and World Wildlife Fund picked up the wolf's cause instead (Mech & Boitani, 2010). By the 1990s, public opinion surveys showed citizens' attitudes supported its protection (Ibid, p. 295).

In the 2000s, a barrage of management actions, lawsuits, and court rulings occurred among interest groups, states, and the federal government over how to manage Great Lakes wolves. These decisions included how to classify distinct population segments, whether state agencies could kill wolves that preyed upon livestock, whether to hunt and trap wolves, and what their ESA status should be (US Fish and Wildlife Service, 2003, 2007, 2008). A 2014 decision ruled that Great Lakes wolves would be relisted as protected under the ESA.

To revive a population of nine remaining wolves on Isle Royale, officials considered and rejected a "genetic rescue" to translocate wolves in 2014. By 2016, an estimated 1,300 moose occupied the island while only a single male–female pair of wolves remained. Those wolves were thought to be highly inbred – father and daughter as well as half-siblings (Peterson & Vucetich, 2016).

As part of an effort to revitalize the wolf population, the NPS hosted four public meetings and solicited feedback for alternative management plans to its Moose-Wolf-Vegetation Management Plan and Environmental Impact Statement (NPS, 2016). This included feedback from all 50 states and other countries. During the public comment period, Minnesota's Grand Portage Band of Lake Superior Chippewa – the closest Native American community to Isle

Royale – opposed translocations. They urged the NPS to wait "at least a decade post-extinction of wolves" to decide whether to restock them, arguing that a delay would provide time to assess other metrics of ecosystem change and see if wolves from the mainland would naturally recolonize the island (Deschampe, 2017). The NPS followed its meetings and feedback system with a draft plan and impact statement for public review, a final plan and statement, and ultimately the decision to translocate 20–30 wolves in an initial three-year period (NPS, 2018).

In September and October 2018, when the NPS released four Minnesota-born wolves (Figure 4.2) in the park, one soon died from pneumonia and another left the island via an ice bridge to return to the mainland. In spring 2019, the NPS released an additional four Canadian wolves, followed by

Figure 4.2 Gray wolf released on Isle Royale National Park. The first release of relocated wolves on Isle Royale was in September 2018.

Source: Photograph by Jacob W. Frank, published with permission of the National Park Service.

another seven. By late March 2019, an estimated 15 wolves occupied Isle Royale (Hoy et al., 2019). By the end of the year, scientists had determined that the new wolves were forming groups that fed, traveled, fought, and bedded together. Necropsies showed that one transplanted female and the last male on the island before the relocations had been killed by other wolves, something the NPS said was "not uncommon as wolves defend and establish their territories and social hierarchy" (NPS, 2019b). The agency later announced that it had confirmed the birth of pups in 2019 and 2020, based on data from GPS collars and images from remote cameras (NPS, 2020).

Media depictions

An important reason for the popularity of Isle Royale animals from a communications perspective is that moose and wolves are charismatic megafauna with cultural meanings beyond their predator–prey relationship. They are objects of public interest, folklore, and media attention (Killion, Melvin, Lindquist, & Carter, 2019; Timmermann & Rodgers, 2005; Williams, Ericsson, & Heberlein, 2002). There is substantial debate about how conservation efforts disproportionately advantage charismatic species at the expense of lesser-known and unglamorous species, especially when resources are limited to conserve or protect species at risk. The moose often engenders empathy or goodwill. It is generally depicted as a large but gentle herbivore standing placidly by a remote wilderness lake, or sauntering along the main street of a small rural town in the northern reaches of the continent.

Public and media perceptions of wolves are more nuanced than those of moose and more complex than the typical human–wildlife conflict depiction (Killion et al., 2019). The wolf ranks among the world's 10 most charismatic animals, yet it is the only one the International Union for Conservation of Nature classifies as of "least concern," meaning it is not a focus of conservation efforts (Courchamp et al., 2018). Examining children's literature, Mitts-Smith (2010) traced the wolf's cultural makeover and conceptualizations, including wolf as predator, as social being, and as canine. That image shift is popular among wolf allies committed to saving the species from extinction and continues to counter traditional rural fears represented in mainstream media.

Sources in environmental reporting

News sources are people, organizations, public agencies, or businesses that provide information to journalists and influence media content in several ways (Messner & Distaso, 2008). Research has found that sources shape narratives by influencing the context in which journalists evaluate an issue, and possibly shape news more than journalists themselves (Berkowitz & Beach, 1993; Reese, Grant, & Danielian, 1994). Source selection can also set news agendas (Salwen, 1995). Research has found that environmental reporting is similar to traditional beats that heavily use government sources, while infrequently including public and environmentalist voices (Lacy & Coulson, 2000; Smith & Norton, 2013). A lack of trust and credibility of environmental reporting has been attributed to this reduction in source diversity, especially when favoring government or industry professionals (Covello, 1989). Similarly, stories about environmental risk with multiple sources have been perceived as more credible and interesting (Cozma, 2006). These perceptions stem from beliefs that environmental protection programs have been mismanaged or officials have made serious errors.

Source diversity is important because environmental policy affects various stakeholders differently. These perspectives can inform the public on diverse environmental impacts, which can potentially influence opinion and policy (Cox, 2013; Smith & Norton, 2013). The predominance of government sources has been attributed to the presence of official spokespeople or public relations staff, making it easier for journalists to find available and suitable sources (Gans, 2004; Smith & Norton, 2013). This is especially true when officials produce press releases and hold press conferences where journalists can quickly garner information and quotes. This can be troublesome for environmental reporting, which already tends to be event-driven, irregular, and too seldom offering in-depth analysis or perspectives of independent sources (Freedman, 2011).

Stories lacking citizen sources can suggest the public has little engagement with the issue and that the media serves as its primary source of information on the topic (Grant & Lawhon, 2014). How sources portray policy influence how a problem is defined; thus, favorable solutions likely align with that problem definition (Rochefort & Cobb, 1993). This is the case when covering wildlife, where management officials dominate the news and stories focus primarily on management actions and conflicts, often about game animals rather than endangered ones (Corbett, 1995). Thus, the majority source type can quickly dominate the media and possibly influence public opinion.

Research questions

RQ1: How did the National Park Service and its partners interact with the news media in connection with the translocation of wolves from the mainland into Isle Royale National Park?

RQ2: How did US and Canadian news organizations cover the translocation of wolves from the mainland into Isle Royale National Park?

Methods

Qualitative analysis

The authors conducted semi-structured, in-depth telephone interviews informed by our research questions in April 2019 with three NPS officials, a Michigan Department of Natural Resources (Michigan DNR) communications director, and a journalist who closely covered the translocations (see Table 4.1). The authors analyzed the interviews by grouping interviewees'

Table 4.1 Semi-structured interviews and date conducted.*

Name	*Title*	*Affiliation*	*Date*
Phyllis Green	Superintendent	Isle Royale National Park	4/3/2019
Liz Valencia	Manager, Interpretation & Cultural Resources Division	Isle Royale National Park	4/3/2019
John Pepin	Deputy Communications Director	Michigan Department of Natural Resources	4/24/2019
Alexandra Picavet	Chief of Communications & Legislative Affairs, Midwest Region	National Park Service	4/30/2019
Tanda Gmiter	Journalist	MLive	4/15/2019

* Table courtesy of Alexander Killion, Mark Neuzil, & Eric Freedman.

answers into categories based on common content to solicit information about the agencies' media strategies and tactics (Saldana, 2009). They also relied on content of press releases and media advisories from Isle Royale National Park and others for additional context.

Quantitative analysis

The content analysis examined news stories printed, broadcast, and posted in the United States and Canada from June to November 2018. It defined news stories as non-advertising matter in a news product and was produced by staff, freelancers, columnists, or correspondents, including wire services and syndicated content. News stories included features and in-depth pieces.

The authors searched five news databases (Newspaper Source Plus, Academic Search Premier, LexisNexis, ProQuest Central, and Google News) for stories with the Isle Royale wolf translocation as the primary topic. To ensure intercoder reliability, the three authors coded a random 10% sample of stories and discussed coding differences until reaching a common understanding (Krippendorff, 2012). Each story included in the final analysis was coded by two authors; a third reader resolved coding inconsistencies for accuracy.

They recorded story attributes including country, news organization, medium, audience, author origin, topic prevalence, points of contention, images, and types of sources. The study defined sources as people, organizations, public agencies, or businesses that provide information to journalists. It also measured the prevalence of topics included in the stories. Topics were listed as major (four or more sentences dedicated to the topic), mentioned, or missing.

The study used Ward's agglomerative hierarchical clustering method to identify common groups of source diversity across stories (Murtagh & Legendre, 2014). It calculated the within-group variance of source-type means for each iteration and chose the fewest number of groups that produced the greatest variance (i.e., groups with distinct patterns of source diversity without over-segmenting the data).

Findings: qualitative results

Planning media coverage

To coordinate communication with the media, the NPS established an Isle Royale National Park Joint Information Center that included representatives of the Michigan and Minnesota DNRs and the US Fish & Wildlife Service. The IRNP superintendent, Phyllis Green, said, "That meant you had a dialogue with the different agencies involved in the project and put key communication personnel in the system." The Michigan DNR deputy communications director, John Pepin, said the Joint Information Center was part of a centralized "incident command structure" with an "incident command team in place." (The Federal Emergency Management Agency uses that type of "regimented structure" for disasters, and the Michigan DNR uses it for wildfires.) Before the translocation decision, a Michigan DNR wildlife biologist handled most of that department's interaction with the NPS. Once the NPS made its decision, the Michigan DNR "looked at it as a Park Service project we were partnering in." Pepin explained:

> In the media realm, having all those people there representing different agencies helps your coordination and reduces misinformation and maximizes resources by co-locating PIOs [public information officers] and the like. We were all, when it could be arranged, in the same place working on stuff. I was privy to drafts of press releases, proofed some drafts, edited some drafts, made some suggestions and was doing photos as well.

Public relations planning started in spring 2018 and ramped up in June 2018 when the NPS prepared to announce its management plan to relocate wolves – "when we realized it was a go," said Liz Valencia, the manager of IRNP's Interpretation & Cultural Resources Division. "The public is really interested in this. They just love it." NPS personnel led the Joint Information Center because other agencies "all knew they wouldn't have convened without us, so they gave us deference," Superintendent Green said, continuing, "Once we were in the actual mode of operation and put in a release to the press, we got reviewed by the [other] agencies. Not all agencies weighed in equally. Those intimately involved did have their communication staff weigh in."

Access for journalists

The NPS and its partners discussed whether to allow journalists at the capture and release sites. Their primary concern was the safety of the wolves and of personnel, such as the veterinarians and capture teams, not what journalists wanted. Alexandra Picavet, the NPS chief of communications and legislative affairs for the Midwest region, said, "We tried to figure out every which-way we could make it work, or at least provide the images and direct access to the people [NPS personnel and other project participants, including a Fish & Wildlife Service photographer] who were there without compromising the project." In addition, government photographers could help if problems arose. "If we said, 'Put your camera down and work with us,' they would," while media photojournalists couldn't be ordered around that way. DNR deputy communications director Pepin cited timing and transportation issues. Allowing journalists on site "was going to be unwieldy, especially if there were a lot of reporters," he said. Superintendent Green also pointed out safety issues during the fall and winter, when most concessions are closed. "How do I get people off if there's a problem? In the winter there's literally no room for error. The odds for survival are slim for people because of response times."

Thus, rather than allowing the press on Isle Royale, the agencies decided to provide photos and videos. As Superintendent Green said, "Once [journalists] find out how tough it is, they say just send them the stuff. We're sending them enough so they're reasonably informed." The IRNP manager, Valencia, said the agency had to be fair and didn't want to decide where to cut off media access: "This was fairly well-received by the press, as long as we posted photos and videos." Superintendent Green noted that the NPS did allow the National Parks of Lake Superior Foundation to send photographers and other people to produce documentary photos and videos "to end up in an educational program relative to the whole program." The foundation, together with the International Wolf Center, raised money to help fund a 2019 relocation (after the study period) when there was insufficient federal funding available. "They have had a few embedded photographers and have gained the trust of the people who work with the animals," Superintendent Green explained.

What did journalists lose by not being on the scene? "I would have loved to talk to the teams that were responsible for the capture," journalist Tanda Gmiter of the MLive newspaper chain said. "That element was missing – how the net trapping went, how difficult it was to spot [wolves] from the air. I really think their angle got missed because I'm sure they have some fascinating stories."

Press releases

Press releases provided the major source of information for journalists. During the study period, IRNP issued nine press releases, plus two media advisories announcing upcoming press conferences. Out of the 11 press releases and advisories, 10 listed Superintendent Green and her phone number as media contacts; IRNP manager Valencia was listed on 9 and NPS official Picavet on 7. Some news organizations used the press releases virtually verbatim. For example, the October 13 *Iron Mountain* (Michigan) *Daily News* article headlined "Four Minnesota wolves moved to Isle Royale" consisted of 13 sentences, all drawn directly from the previous day's press release; the newspaper changed only one word, replacing "today" with "Friday," and did not disclose that it was a press release.

There were some media-related policy disagreements within the communications team. One concerned whether to treat all journalists the same. "One thing we tend not to do in the federal government is give exclusives," NPS official Picavet explained. "The state entities, the [Michigan] DNR, and [the] Minnesota [DNR] had different constraints but understood not giving exclusives, but having equal access." However, the Grand Portage Band representative who worked with the media "had a very different view" and would have provided exclusives if the band "had its druthers." She said, "It could be described as different tactics."

Michigan DNR's Pepin pointed to another disagreement with a biologist representing the Grand Portage Band:

> When one of the wolves we were going to translocate died [on the mainland on September 27, 2018, after being trapped and sedated], we put that out to the media. We wanted everything to be transparent and available for the media.

The tribal representative "was not happy" with that action, however, and "felt it made the biologist team look bad. He felt the fatality was something that's not necessarily expected but can happen when trapping wolves. He was upset the Park Service decided to go ahead and release that information." A necropsy later determined that anesthesia-related stress was the likely cause of death.

The wolf's death was disclosed in the third through fifth paragraphs of a release titled "Isle Royale Wolf Relocation Update;" the first two paragraphs announced the capture of four wolves on the Grand Portage Reservation.

In retrospect, would the communications team have done anything differently? "It was the very first operation, so we had no idea what to expect. There were too many unknowns," NPS official Picavet said. The 35-day US government shutdown from December 22, 2018, until January 25, 2019 (after the study period), threw "a great big monkey wrench" in the process.

The NPS media strategy, coupled with lack of physical access for journalists, enabled the agencies to largely control and shape press coverage. That is reflected by the close similarity in stories by print, online, and broadcast news outlets, the commonality of much of the data in their stories, and the dominance of a limited number of NPS officials – most often Superintendent Green and Mark Romanski, the park's chief of natural resources, as named sources with quotes taken from press releases and interviews.

Assessments

Officials sounded satisfied with the press coverage and expressed only minor complaints. Superintendent Green described coverage as "mostly accurate" and said, "The stories were more accurate when we had total control over them." She added, "The toughest thing for our press people to deal with was one fatality in the fall. We were [just] going to summarize the operations but we knew there was a possibility it [a death] would happen."

Pepin, the Michigan DNR deputy communications director, said, "We did not post images of the wolf that died, so the media was using images of ones that survived." The NPS's Picavet said she saw a "couple of stories . . . that [led communications and agency employees to wonder] 'what are they talking about here?'" Overall, however, she said, "With the information we had, money constraints, time constraints, the overwhelming concern was making sure it was a successful project, I believe through what I've seen of the coverage there's been really full coverage, with a lot of photography with direct access to the superintendent and other key players."

What was MLive journalist Gmiter's assessment of the Joint Information Center? "I'd give an overall grade of A-minus to B-plus," and she praised the NPS for candor, although adding that "some of the information has been a little slow to come out." In her opinion, "the National Park Service has done a good job getting out information that isn't fun to tell" – such as wolf deaths and the fact one wolf used an ice bridge to leave the island.

Findings: quantitative results

To analyze how the news media covered the translocation, the content analysis initially produced 366 articles; after removing those that did not meet the inclusion criteria (primarily because they were not about the Isle Royale wolves), 68 articles were coded. Of them, 63 stories (93%) were from US media and five (7%) from Canadian media. The vast majority (54, or 79%) came from newspapers, magazines, and wire services (in print and/or online), rather than radio (18%) or television (3%). Among the stories, 13 (19%) included points of contention, most often concerning whether the NPS correctly decided to relocate wolves.

Expert sources from the US government were most common, with 76% of articles having at least one and 25% having at least two. Academics appeared in 25% of articles and environmental interest groups in 15%. However, rarely (6%) were all three of these sources combined in an article. Similarly, only 3% of the articles included issue experts from both the US and Canadian governments. There was also sparse representation from tribal (7%), Canadian government (4%), and ordinary sources (4%).

The most commonly discussed topic (79%) was information on logistics and implementation of the translocation, followed by the translocation decision (35%) and Isle Royale ecology (35%). These three topics commonly co-occurred across stories. Economic, tourism, or business impacts were never major topics. Most articles (91%) included at least one visual element, primarily photos from press releases but occasionally a map, graph, or embedded video. Photos with wolves (78%) far outnumbered photos with moose (13%). An occasional photo featured both (12%), for example, a wolf feeding on a moose carcass.

The Ward's agglomerative hierarchical clustering model resulted in four groups of source diversity, ordered here by amount of diversity (Figure 4.3). The first group, called the Diverse Source Group (22 stories), had the highest prevalence of US government sources and the most source diversity. This group included eight of the 13 stories with points of contention. The second group, the Academic Source Group (12 stories), was dominated by academic sources, with US government sources the second largest in the group. The third group, the US Government Source Group (23 stories), included *only* US government sources. The fourth group had no named sources (No Source Group; 11 stories); these stories did not quote a source or include/attribute factual information from somewhere else.

Each story was also classified by the media outlet's primary type of audience. Stories in the Diverse Source Group were most evenly distributed across all audience types while the stories in the Academic Source Group were primarily featured by media outlets aimed at science or outdoor audiences (Figure 4.4). The majority of the stories in the US Government Source Group were directed at a national audience, and the No Source Group was featured most in regional stories.

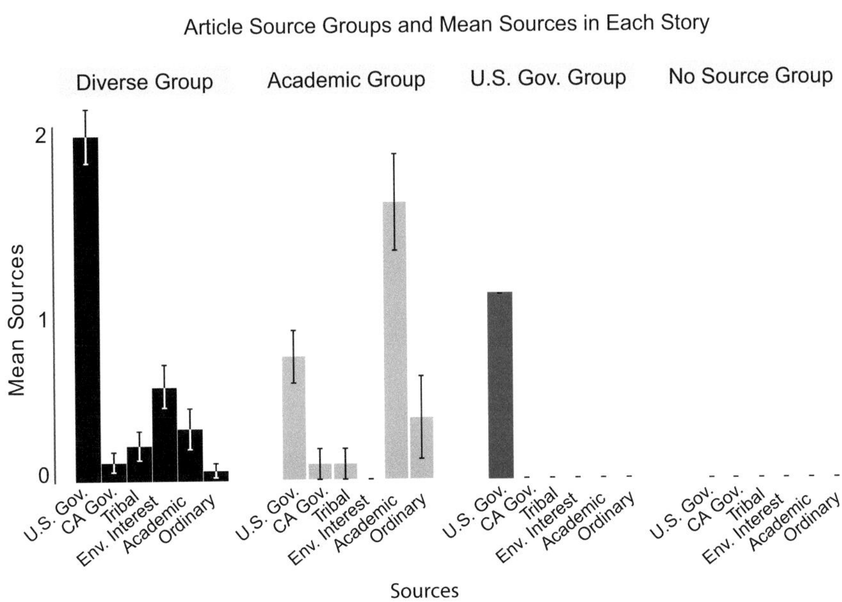

Figure 4.3 Four source groups determined by the Ward Hierarchical Clustering model. Bars represent the mean number of source types included in each story for each source group. Error bars represent the standard deviation of the number of source types in each story.

Source: Graph courtesy of Alexander Killion, Mark Neuzil, and Eric Freedman.

Discussion and conclusions

Newsmakers and journalists recognize the importance of press releases as sources of information. Press releases are not neutral because issuers seek to present information, opinion, and analysis in ways that support their actions, decisions, products, and points of view. Meanwhile, continually shrinking newsrooms and dwindling numbers of reporters, editors, photographers, and other news professionals leave the remaining staffers increasingly dependent on press releases for story ideas and content. This often occurs without additional independent reporting to verify assertions and to seek contrary perspectives and data. As Rodman notes, "Today, much of the news we receive from any medium consists of information supplied by press releases. Newspapers especially rely on them" (2012, p. 349). Yet, journalist Gmiter said, "Stories not from press releases resonate more and stay longer than ones that other news outlets are writing" based on press releases.

This study shows that was true for coverage of the translocations. Journalists used large portions of the information from press releases, had no alternative sources for up-to-date photos and video, and heavily relied on the same sources proffered by the NPS and the Joint Information Center for additional quotes or data. Without physical access to locations where news was being made (capture and release sites) and unable to interview project personnel on the ground, journalists relied on the NPS, which largely drove what was reported and what was not.

Stories comprised four well-defined groups of sources. Although the Diverse Source Group (32% of stories) included many source types, the stories always had at least twice the number of US government sources as other types. Additionally, the most common source group, the US Government Source Group (34% of stories), included only a single US government source. Sources from the Canadian government were most infrequently used, which is surprising given the island's proximity to Canada, the fact that some wolves came from Canada, and one translocated wolf walked back to Canada.

Stories for a national audience most often included only a US government source (Figure 4.4), while regional or science audiences were more likely to encounter greater source diversity. This difference may be attributed to local journalists having greater access to diverse sources, and their readers being more interested in an assortment of perspectives affecting stakeholders, while national stories relied more on press releases for information. The Academic Source Group was more likely to be used for a science or outdoor audience, suggesting these readers received stories that were informed by independent scientific experts.

Another indicator of journalistic dependency on press releases is the virtual absence of "ordinary" people as news sources, such as park users or tourism and business representatives on the nearby Minnesota and Michigan mainland. This is true despite high public interest in Isle Royale among past and prospective visitors and among mainland businesses serving those visitors. Similarly, the

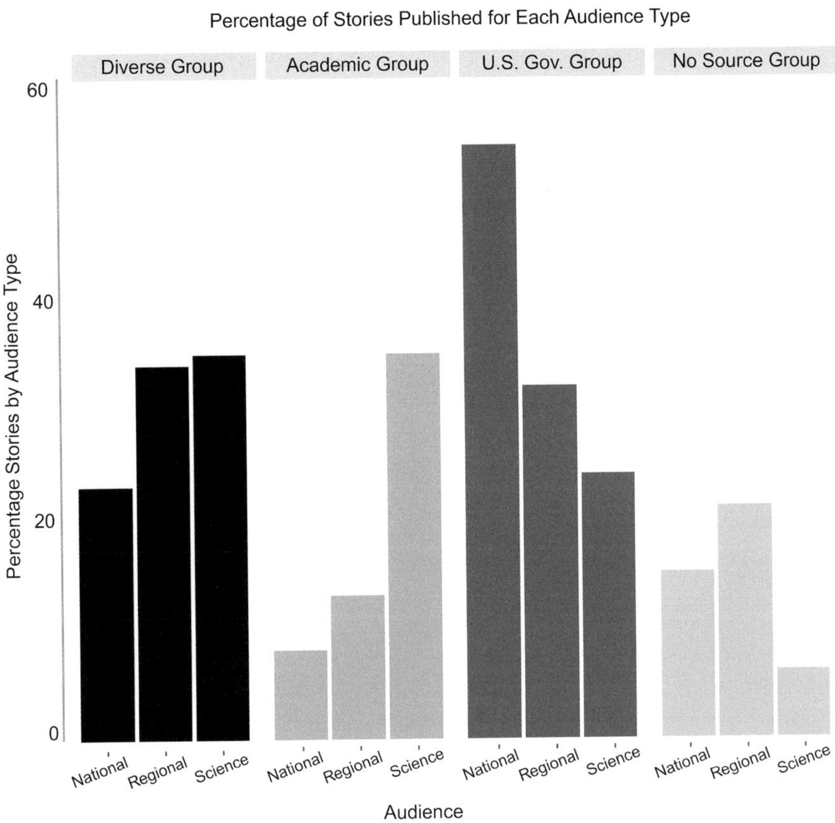

Figure 4.4 Percentage of total stories published for each audience type, separated by source groups. Audience type is defined as the media outlet's primary type of audience. The characteristics of these four source groups are shown in Figure 4.3.

Source: Graph courtesy of Alexander Killion, Mark Neuzil, and Eric Freedman.

tribal voice of the Grand Portage Band – a source of some of the translocated wolves – was rarely heard, although many IRNP press releases mentioned its participation in the project. For the NPS and partners, the Joint Information Center's media strategy and tactics proved successful in telling their stories and appearing transparent on a topic of great public and press interest. At the same time, they made paramount the conservation management goals of the project and the safety of its participants.

By controlling access to locations where news occurs, authorities infringe on reportorial independence and the ability of journalists to gather information through the exercise of their own professional judgment. On the other hand, it is not unreasonable to deny journalists access to disaster areas, no matter how newsworthy the events are and no matter how contentious such restrictions may be (Muller & Gawenda, 2010). In the Isle Royale situation, journalists did not seriously question the legitimacy of NPS concerns for the welfare of the wolves and of the personnel, even if they felt their presence would not impair those concerns.

That said, such restrictions do influence what the public gets to read, see, and hear from independent journalists. Denial of access is especially damaging to visual journalists – photographers and videographers. "Through photography we witness time frozen through the mental lenses of the image maker" (Bossen, Freedman, & Mianecki, 2013, p. 5). Without access to journalists' independent reporting and observation, audiences saw the Isle Royale translocations through only the mental and physical lenses of official government image makers.

Superintendent Green said the NPS will continue to work with journalists. "We're way more amenable in the summer because they can get there. If they're doing a news story, we can accommodate them." She cautioned there is no assurance journalists will spot any wolves because they generally stay away from people. "You don't see wolves on Isle Royale. It's so rare."

The authors of this chapter, environmental journalists and a conservation scientist who focuses on environmental problems, are concerned with how messages are communicated and how access restrictions affect the portrayal of resource management actions. Remote or restricted areas like Isle Royale are commonplace in environmental studies and reporting. Whether due to the lack of human presence or social conflict that endangers scientists and journalists, these places are highly affected by management decisions. It is the responsibility of journalists and scientists to ensure that coverage includes diverse perspectives.

Acknowledgments

This chapter draws on research previously presented by Killion, A., Neuzil, M., & Freedman, E. (2019). *The wolves of fate: Media coverage of the 2018 Isle Royale "genetic rescue."* International Environmental Communication Association, Conference on Communication and Environment.

References

Berkowitz, D., & Beach, D. W. (1993). News sources and news context: The effect of routine news, conflict and proximity. *Journalism Quarterly*, *70*(1), 4–12.

Bossen, H., Freedman, E., & Mianecki, J. (2013). Hot metal, cold reality: Photographers' access to steel mills. *Visual Communication Quarterly*, *20*(1), 4–19.

Cole, J. E. (1957). *Isle Royale wolf investigations, winter of 1956–57: Isle Royale National Park files*. Special Report. Houghton, MI: Michigan Technological University.

Corbett, J. B. (1995). When wildlife make the news: An analysis of rural and urban North-Central US newspapers. *Public Understanding of Science*, *4*(4), 397–410.

Courchamp, F., Jaric, I., Albert, C., Meinard, Y., Ripple, W. J., & Chapron, G. (2018). The paradoxical extinction of the most charismatic animals. *PLoS Biology*, *16*(4), e2003997.

Covello, V. T. (1989). Informing people about risks from chemicals, radiation, and other toxic substances: A review of obstacles to public understanding and effective risk communication. *Prospects and Problems in Risk Communication*, 1–49.

Cox, R. (2013). *Environmental communication and the public sphere*. Thousand Oaks, CA: Sage.

Cozma, R. (2006). Source diversity increases credibility of risk stories. *Newspaper Research Journal*, *27*(3), 8–21.

Deschampe, N. W. (2017). Letter from the Grand Portage Reservation Tribal Council to the superintendent of Isle Royale National Park. Houghton, MI: Michigan Technological University.

Freedman, E. (2011). Environmental journalism in Kyrgyzstan and Kazakhstan: Reporting scarce amid environmental and media problems. *Applied Environmental Education & Communication*, *10*(2), 126–134.

Gans, H. J. (2004). *Deciding what's news: A study of CBS Evening News, NBC Nightly News, Newsweek, and Time*. Evanston, IL: Northwestern University Press.

Grant, S., & Lawhon, M. (2014). Reporting on rhinos: Analysis of the newspaper coverage of rhino poaching. *Southern African Journal of Environmental Education*, *30*(0), 39–52.

Grooms, S. (2002). A brief history of wolf research. *International Wolf*, *21*(2), 9.

Hoy, S. R., Peterson, R. O., & Vucetich, J. A. (2019). Ecological studies of wolves on Isle Royale: Annual report 2018–2019. Retrieved from https://isleroyalewolf.org/sites/default/files/annual-report-pdf/annualrep2016webversion.pdf

Killion, A. K., Melvin, T., Lindquist, E., & Carter, N. H. (2019). Tracking a half century of media reporting on gray wolves. *Conservation Biology*, *33*(3), 645–654. doi:10.1111/cobi.13225

Krippendorff, K. (2012). *Content analysis: An introduction to its methodology*. Thousand Oaks, CA: Sage.

Lacy, S., & Coulson, D. C. (2000). Comparative case study: Newspaper source use on the environmental beat. *Newspaper Research Journal*, *21*(1), 13–25.

Line, L. (1996, March 19). In long-running wolf-moose drama, wolves recover from disaster. *New York Times*.

Mech, L. D., & Boitani, L. (Eds.). (2010). *Wolves: Behavior, ecology, and conservation*. Chicago: University of Chicago Press.

Mech, L. D., & Breining, G. (2020). *Wolf Island: Discovering the secrets of a mythic animal*. Minneapolis, MN: University of Minnesota Press.

Messner, M., & Distaso, M. W. (2008). The source cycle: How traditional media and weblogs use each other as sources. *Journalism Studies*, *9*(3), 447–463.

Mitts-Smith, D. (2010). *Picturing the wolf in children's literature*. New York: Routledge.

Muller, D., & Gawenda, M. (2010). Ethical free-for-all over media access to the fire zone. *Media International Australia*, *137*(1), 71–79.

Murtagh, F., & Legendre, P. (2014). Ward's hierarchical agglomerative clustering method: Which algorithms implement Ward's criterion? *Journal of Classification*, *31*(3), 274–295.

National Park Service. (2016). Isle Royale wolves EIS public scoping newsletter. Retrieved from https://isleroyalewolf.org/sites/default/files/Revised scope of Isle Royale Wolves EIS.pdf

National Park Service. (2018). Final environmental impact statement to address the presence of wolves at Isle Royale National Park. Houghton, MI: National Park Service.

National Park Service. (2019a). Park statistics. Retrieved from https://irma.nps.gov/STATS/

National Park Service. (2019b). Group formation of new Isle Royale wolves leads to territorial aggression. Retrieved from www.nps.gov/isro/learn/news/group-formation-of-new-isle-royale-wolves-leads-to-territorial-aggression.htm

National Park Service. (2020). Wolf pups born on Isle Royale. Retrieved from www.nps.gov/isro/learn/news/wolf-pups-born-on-isle-royale.htm

Peterson, R. O., & Vucetich, J. A. (2016). Ecological studies of wolves on Isle Royale: Annual report 2015–2016. Retrieved from https://isleroyalewolf.org/sites/default/files/annual-report-pdf/annualrep2016webversion.pdf

Reese, S. D., Grant, A., & Danielian, L. H. (1994). The structure of news sources on television: A network analysis of "CBS News," "Nightline," "MacNeil/Lehrer," and "This Week with David Brinkley." *Journal of Communication, 44*(2), 84–107.

Rochefort, D. A., & Cobb, R. W. (1993). Problem definition, agenda access, and policy choice. *Policy Studies Journal, 21*(1), 56–71.

Rodman, G. (2012). *Mass media in a changing world.* New York: McGraw-Hill.

Saldana, J. (2009). An introduction to codes and coding. *The Coding Manual for Qualitative Researchers, 3.*

Salwen, M. B. (1995). News of Hurricane Andrew: The agenda of sources and the sources' agendas. *Journalism & Mass Communication Quarterly, 72*(4), 826–840.

Smith, H. M., & Norton, T. (2013). Environmental groups on par with government sources. *Newspaper Research Journal, 34*(1), 50–61.

Timmermann, H. R., & Rodgers, A. R. (2005). Moose: Competing and complementary values. *Alces, 41*, 85–120.

US Fish and Wildlife Service. (2003). Final rule to reclassify and remove the gray wolf from the list of endangered and threatened wildlife in portions of the conterminous United States. Washington, DC: US Fish and Wildlife Service.

US Fish and Wildlife Service. (2007). Removing the Western Great Lakes distinct population segment of the gray wolf from the list of endangered and threatened wildlife. Washington, DC: US Fish and Wildlife Service.

US Fish and Wildlife Service. (2008). Reinstatement of protections for the gray wolf in the Western Great Lakes and Northern Rocky Mountains in compliance with court orders. Washington, DC: US Fish and Wildlife Service.

US Fish and Wildlife Service. (2011). Gray wolf recovery in Minnesota, Wisconsin and Michigan. Washington, DC: US Fish and Wildlife Service.

Williams, C. K., Ericsson, G., & Heberlein, T. (2002). A quantitative summary of attitudes toward wolves and their reintroduction (1972–2000). *Wildlife Society Bulletin, 30*(2), 575–584.

5 How African is the African elephant?

Reflections on elephant poaching and conservation in African and UK newspapers

Harriet Croome and Clare Wilkinson

In 1979, 1.3 million elephants roamed Africa, falling to approximately 600,000 by 1989 – a result of unprecedented ivory poaching (Douglas-Hamilton, 1987; van Aarde, Jackson, & Ferreira, 2006). In response, the Convention on International Trade in Endangered Species of Wild Flora and Fauna (CITES) granted African elephants the highest level of protection under Appendix I, prohibiting international trade in them and their by-products. At every CITES Conference of the Parties since, the effectiveness of the 1989 listing has sparked debate and, despite continued conservation efforts and legislative measures, elephants remain imperiled by poaching, human–elephant conflict, and habitat loss (Stiles, 2004). Approximately 415,000 remain (see Figure 5.1), declining on average by 8% per year (Chase & Landen, 2019; Chase et al., 2016).

Elephant conservation and human-elephant conflict mitigation initiatives have been criticized for failing to integrate local people's knowledge and interests, and recognize that success may come from accounting for cultural values and socioeconomic concerns that shape the lives of people living alongside elephants (Shaffer, Khadka, Van Den Hoek, & Naithani, 2019).

Since mass media coverage is one place where diverse viewpoints converge to impact conservation policy making and public opinion, this chapter examines representations of elephant conservation in newspapers. It draws on a content analysis of English-language articles published in the United Kingdom and 11 African elephant range states between 1996 and 2017, and semi-structured interviews with stakeholders. The research explores whether newspaper reporting mirrors an understatement of local knowledge and actors in African elephant conservation and considers how this may influence conservation itself.

Statement of the problem

Elephants mean different things to different people: a symbol of wonder, kinship, and sensibility for some; fear, insecurity, and loss of livelihood for others. Few studies, however, have examined how local and external knowledge are presented, co-operate, or lock horns in public discourse, or have discussed its potential impacts on conservation practice and policy making.

Figure 5.1 An African elephant (*Loxodonta africana*), the largest land mammal in the world.
Source: Photograph by Michelle Gadd/US Fish & Wildlife Service.

This chapter asks: How do newspapers portray elephants? It considers how and why journalists select sources, how sources acquire legitimacy, and relationships among sources, frames, and the cultural characteristics of the place of publication. The sources selected are important. In an analysis of the "Save the Serengeti" Facebook campaign, Gardner (2017) demonstrates that an international perspective propagated by online campaigners drove decisions to manage the greater Serengeti ecosystem. That perspective helped justify land dispossession and eviction of local people who were unable to participate online. In this example, greater stakeholder inclusivity in reporting may have engendered more democratic policy making, but little research exists into how journalists legitimize vulnerable, marginal, and local voices (Hivon, Lehoux, Denis, & Rock, 2010).

Globally, falling advertising revenue and circulation are putting media outlets under economic strain (Rubado & Jennings, 2019). Journalists are increasingly desk-bound and time-poor, leading to greater reliance on internet resources and pre-packaged news stories (Vicente, 2017). Researching climate change coverage in Africa, Tagbo (2010) observes that the bulk of African newspaper content was culled from foreign institutions and researchers owing to time pressures, resulting in coverage reflecting Western rather than African perspectives and values.

Western involvement in saving African wildlife, reminiscent of colonial exclusionary practices arguably continues to color African conservation to the detriment of Indigenous livelihoods (Dowie, 2011). Mbaria and Ogada (2016, p. 104), crystallizing resentment towards the dismissal of local knowledge in favor of Western expertise, state: "Kenya does not need experts nurtured on the European education system. . . . [F]ormally uneducated men and women, whose cognitive abilities remain broadly unacknowledged . . . provide solutions the West can never offer to Africa."

Mass media constitute a significant arena in which scientific and environmental issues come to the attention of decision-makers, interest groups, and publics (Shoemaker & Reese, 1996). How the media present, or frame, an issue therefore shapes the definition, trajectory, and resolution of policy issues (Gamson & Modigliani, 1989; Nisbet, Brossard, & Kroepsch, 2003). Framing organizes discourse according to certain perspectives and involves selection, meaning inclusion or exclusion of fact, and hierarchical arrangement of elements to produce meaning (Entman, 1993). This informs audience understanding, enabling attachment of different degrees of importance to different claims and claims-makers. Sources and perspectives that journalists choose to emphasize because of their own value systems, or the perceived values of their audiences, will likely be considered important by readers also (Gamson & Modigliani, 1989; Entman, 1993).

Research shows a "reluctance of mainstream corporate and commercial media to critically address major environmental and conservation issues" (Blewitt, 2011, p. 719). The result is a public discourse that emphasizes and reproduces familiar Western ideologies around conservation practice. When it comes to conservation and the environment, the media is credited with an "unrealized critical educative potential," so the nature of reporting may fundamentally impact policy solutions (Blewitt, 2011, p. 719; Soroka, Lawlor, Farnsworth, & Young, 2012). Examining media and Indigenous policy making in Australia, McCallum and Waller (2013) identified a direct link between the "internalized media-logic" of policy makers and press announcements being turned into policy. "Media-logic" describes an increasing societal reliance on media through which policy makers internalize and replicate media routines, practices, and priorities. These findings highlight the importance of including a plurality of voices and viewpoints in media coverage of science and the environment to advance representative decision-making.

Research question

Increasingly, researchers are urged to pay attention to sociological concerns regarding power and inequality in public communications and to identify how economic, political, and cultural power determine one's ability to participate in and influence environmental reporting (Hansen, 2011). But how can we know what motivates coverage of elephant conservation if we do not hear from journalists? How can we know whether relations among journalists and certain stakeholders are problematic if we do not know what those relations look like?

Studies adopting a social constructionist theoretical framework have made headway into understanding dynamics among source, journalist, and place. Researching representations of human–elephant conflict in Indian and international media, Barua (2010) found that elephants were less likely to be vilified in Indian media due to their revered cultural position. Animal rights, however, were discussed only in international reporting, employed as a strategy to gain sympathy and funding from international publics whose cultural value system resonates with these issues (Barua, 2010).

Similarly, Tyrrell and Clark (2014) found that no newspapers outside Canada reported the potential impact of an uplisting of polar bears on Inuit livelihoods; instead, they privileged animal welfare and extinction narratives. Journalists and animal rights organizations providing the primary source base for their study knew what would resonate with their audiences – and what resonates, sells.

Thus the research question: Does the privileging or omission of certain voices and frames in newspapers affect the extent to which African elephant conservation can be implemented locally?

Method

Conservation of African elephants is highly complex, intersecting science, human development, global commercial interests, and cultural belief systems, and involving many stakeholders. This study, therefore, adopted a mixed-methods, cross-continental approach, combining quantitative content analysis with qualitative semi-structured interviews.

A Nexis database search between April 30, 1996, and April 30, 2017, using search terms "elephant" and "poaching or conservation" generated 283 English-language articles. A sample covering 21 years would allow trends over time and impacts of significant events to become evident. The sample is sufficiently post-apartheid for press censorship to have diminished in South Africa. Significant events in the period include the downlisting of key elephant populations to CITES Appendix II in 1997 and 2000, high-profile ivory stockpile destructions, and population censuses.

Removing irrelevant articles left a sample of 161 articles from the United Kingdom and 11 African nations: South Africa, Botswana, Namibia, Kenya, Tanzania, Zimbabwe, Rwanda, Uganda, Cameroon, Zambia, and Mozambique. A larger sample size was beyond the scope of this study, so the United Kingdom is used as a proxy for reporting from the West/Global North. It is possible that English-language reporting outside of the United Kingdom may differ, and so too may English-language and non-English-language reporting within African nations. For example, Bhatia, Athreya, Grenyer, and MacDonald (2013) identified differences between English-language and non-English-language reporting of human–leopard conflict in Mumbai. They suggested the socioeconomic status and preferences of English-language and non-English-language journalists correlated with a propensity to attribute fault to either people or leopards, respectively.

The authors analyzed each article using a coding schedule, accompanied by an instructional codebook, produced inductively using a pilot sample of every tenth article (Riffe, Lacy, & Fico, 2005). Variables included length, date, and type of article; continent and country of publication; number, type, and order of sources cited; type and order of frames present; celebrity presence; and "African voice." This study defined African voice as "an individual who is Africa-born and lives/works predominantly in Africa." It made no distinction among Africans of any ethnicity. To distinguish between African and non-African voices is not to say that the two are necessarily opposed to one another or that they are two homogeneous groups (Nyamnjoh, 2017). Rather, it is one admittedly imperfect way of assessing whether a plurality of perspectives, configured around one's status as local or external to Africa, exists in reporting on elephant conservation.

The authors created a typology of discourse frames (see Table 5.1) by adapting existing analyses of environmental media coverage to encompass key elements

Table 5.1 A framing typology for representations of African elephants in UK and African newspapers.*

Frame	*Description*
Information	Highlights population decline through statistical evidence. Focus on poaching, numbers of animals killed, or amount of ivory seized. Neutral and scientific in tone.
Crime/Law Enforcement	Focus on poachers rather than elephants. Highlights criminality, corruption, danger to human lives, involvement of criminal syndicates, law enforcement, and sentencing.
Animal Welfare	Focus on animal rights and welfare. Presents saving elephants as an end in itself. Mentions extinction risk, preservation for future generations. Biased towards elephants over people.
International Aid/ Partnerships	Global narrative. Highlights perceived need for foreign funding and expertise.
Conservation for Development	Focus on links between conservation and sustainable livelihoods, commodification of natural resources for socioeconomic development. Communities may benefit, such as through jobs creation, but do not necessarily drive conservation.
Community-Based Conservation	Focus on community-based conservation with communities at front and center of conservation policy and practice. Highlights sustainable natural resource use, African solutions, and Indigenous expertise.
Ivory Demand	Focus on consumer demand for ivory.
Solutions/Ivory Management	Focus on solutions to poaching crisis. Includes discussion of ivory stockpile destruction, sales banning/legalization, and policy implementation.
Technology	Highlights role of novel technologies and management practices. May also highlight increasing sophistication of poachers' weaponry.
Human–Elephant Conflict	Focus on HEC, degradation of habitat, loss of lives and livestock, difficulties of living alongside elephants.

* Table courtesy of Harriet Croome & Clare Wilkinson.

and recurring themes (Barua, 2010; Nisbet et al., 2003). Taking title frame as an example, an article titled "David Attenborough calls on ordinary people to help save Africa's elephants" was assigned to the international aid frame, while "Deal to halt illegal ivory trade signed" was assigned to the solutions and ivory management frame. Each item was coded for headline, lead, and overall dominant frame to establish the degree of significance and credibility that journalists attributed to each frame or source. This arrangement of news values is likely to be routine practice for journalists and indicates what they deem most interesting, relevant, or important to their audiences (Anderson, Allan, Peterson, & Wilkinson, 2005).

Combining content analysis with stakeholder interviews enabled the study to go beyond purely descriptive conclusions to interpret media representations in social, cultural, and professional contexts (Carvalho, 2008; Kwansah-Aidoo, 2001). A semi-structured approach ensured all questions could be covered in one interview and that data was comparative. Interviewees were relevant media, policy, and conservation stakeholders. Of 27 individuals and organizations contacted, 11 agreed to participate. Interviews lasted between 23 and 55 minutes and took place between May 20, 2017, and September 13, 2017, in person or via Skype or WhatsApp, and were audio-recorded and transcribed (see Table 5.2). The authors assigned the transcribed data to common themes, or nodes, using NVivo 11 (Abell & Myers, 2008).

Findings

Source analysis

The 161 articles contained 381 sources. The minimum number of sources in an article was zero, the maximum nine, and the mean, 2.68. Table 5.3 details type and frequency of sources cited. The most frequently cited sources were *International Conservation Organizations* (32%; n = 122), *Government/Ministerial* (21.8%; n = 83), and *Academic/Scientific* (14.2%; n = 54). *International Funders* (2.1%; n = 8) appeared more often than *Local/Community Representatives* (1.3%; n = 5) and in contrast to their more dominant international counterparts, *African Conservation Organizations* accounted for only 3.4% (n = 13) of sources.

Across the dataset, 43.3% of lead sources represented "African voice" and 25.5% did not; 31.2% were classified as "not applicable" because the source did not represent an individual voice. For the second source, 22.8% were coded "yes," 38.5% "no," and 57.7% "N/A." In African reporting alone, African voices accounted for the most lead sources (48.4%; non-African 18.3%; N/A 33.3%). In news articles from UK media outlets, non-African voices received the greatest share (38.3%; African 29.8%; N/A 31.9%). After aggregation, "African voice" was either a lead or second source in 34.6% of articles. Just 1.1% of these were Local/Community Representatives, and only 3% were African Conservation Organizations.

Table 5.2 Interviewees*

Interviewee Name	*Job Title/Description*	*Organization*	*Date*	*Method*	*Length (Minutes)*
Rosie Cooney	Head of Sustainable Use and Livelihoods Group	IUCN	7/10/17	Skype Audio	55
Anonymous 1	Senior reporter – based in Zimbabwe	Zimbabwean Newspaper	8/24/17	WhatsApp Audio	23
Lamson Malukele	Community Conservancy Leader	Makuleke Community Conservancy, South Africa	9/13/17	WhatsApp Audio	29
Karl Mathiesen	Freelance journalist – based in the United Kingdom	Writing for *The Guardian*	8/22/17	In person. London, United Kingdom	53
Johnny Rodrigues	Director	Zimbabwe Conservation Task Force	7/10/17	Skype Video	54
Kaddu Sebunya	President	African Wildlife Foundation	8/25/17	Skype Audio	43
Daniel Stiles	Independent Ivory Trade Consultant	N/A	8/1/17	Skype Video	39
Julian Sturgeon	Founder	Resources Trust Africa	8/18/17	Skype Audio	50
Richard Thomas	Global Comms. Coordinator – based in the United Kingdom	TRAFFIC	8/14/17	Skype Video	43
Clive Walker	Founder Author and correspondent – based in South Africa.	Endangered Wildlife Trust	5/20/17	In person. Vaalwater, South Africa	43
Anonymous 2	N/A	N/A	8/18/17	Written Response via Email	N/A

* Table courtesy of Harriet Croome & Clare Wilkinson. Information correct as of 2017.

Across African reporting, 6% of stories included celebrities, while 23% of UK articles included celebrities. In African countries with comparable sample sizes, South Africa and Kenya, the figures were 5% (n = 2) and 17% (n = 5), respectively. Only one celebrity mentioned, Kenyan musician Julius Owini, is Africa-born. All others were white Westerners. Notably, in articles including celebrities, the "animal welfare" and "international aid/partnerships" frames were dominant, while "conservation for development" and "community-based conservation" frames were absent.

Table 5.3 Type and frequency of sources.*

	Dominant Source		*As First Source*		*As Second Source*	
Academic/Scientific	54	14.2%	18	11.50%	18	15.80%
African Conservation Organization	13	3.4%	4	2.50%	3	2.60%
Celebrity	12	3.1%	6	3.80%	5	4.40%
CITES	13	3.4%	7	4.50%	2	1.80%
International Conservation Organization	122	32%	43	27.30%	45	39.50%
International Funder	8	2.1%	2	1.30%	6	5.30%
Local/Community Representative	5	1.3%	1	0.70%	2	1.80%
National Parks and Wildlife Service	43	11.3%	22	14%	9	7.90%
Government/ Ministerial	83	21.8%	45	28.70%	19	16.70%
Safari/Tourism Professional	13	3.4%	3	1.20%	2	1.80%
Other	15	3.9%	6	3.80%	3	2.60%

* Table courtesy of Harriet Croome & Clare Wilkinson.

Frame analysis

The most commonly employed frames were "information" in 95% of articles, "crime/law enforcement" in 63%, "ivory demand" in 53%, and "international aid/partnerships" in 51%. Table 5.4 details the use and frequency of all frames.

Significant differences between framing in UK and African coverage are apparent. African articles were more likely to include the "community-based conservation" frame (39%) than UK articles (11%). The same was true of "conservation for development," with 44% of African articles employing the frame, in contrast to 30% of UK articles. In contrast, UK reporting was proportionately more likely to use "animal welfare" (United Kingdom 60%; Africa 35%) and "technology" frames (United Kingdom 23%; Africa 18%). UK reporting was also more likely to include the "human–elephant conflict" frame (United Kingdom 23%; Africa 14%), running contrary to geographic-distance theory.

In both Africa and the United Kingdom, the dominant overall frame was "information" (United Kingdom 28%; Africa 24%). In African reporting, "crime/law enforcement" was the next most dominant (19%); in UK articles it was "international aid/partnerships" (21%). Although the "animal welfare" frame did not feature highly among dominant overall frames, it was the dominant headline (49%) and opening frame (38%) in UK reporting, and the

Table 5.4 Presence of frames across reporting.*

Frame	All Articles		United Kingdom		Africa	
Information	153	95%	45	96%	108	95%
Crime/Law Enforcement	102	63%	26	55%	76	67%
Animal Welfare	68	42%	28	60%	40	35%
International Aid/Partnerships	82	51%	27	57%	55	48%
Conservation for Development	64	40%	14	30%	50	44%
Community-Based Conservation	50	31%	5	11%	45	39%
Ivory Demand	85	53%	28	60%	57	50%
Solutions/Ivory Management	67	42%	29	43%	47	41%
Technology	32	20%	11	23%	21	18%
Human–Elephant Conflict	27	17%	11	23%	16	14%

*Table courtesy of Harriet Croome & Clare Wilkinson.

dominant opening (28%) and headline (22%) frame after "information" by a margin of one article in African reporting.

The "information," "crime/law enforcement," and "international aid/partnerships" frames were consistently dominant as overall frames across the time period. Headline frames became inclined towards "animal welfare" over time, rising from 17% of reporting in 2010–2012 to 31% in 2016–2017. The emergence of the "community-based conservation" frame is also notable. Between 1996 and 2009, "community-based conservation" did not feature as a dominant frame. It featured once as a dominant frame in 2010–2012, eight times in 2013–2015 (four in headlines), and twice in 2016–2017 (once in a headline).

Qualitative semi-structured interview analysis

The authors did not share the results of the content analysis with interviewees. Instead, they asked open-ended questions on related themes; for example, what element of the elephant population crisis and conservation do you think is most prevalent in the media? How did you go about finding the information/sources for this article? This section examines four recurrent themes from the interviews: source and narrative imbalance; source–journalist interactions; barriers to diversity; and media–policy interactions.

Source and narrative imbalance

Interviewee comments centered on the dominance of simple, events-based reporting and the near-exclusion of more complex narratives. The former head of the International Union for Conservation of Nature Sustainable Use and Livelihoods Specialist Group (IUCN SULi), Rosie Cooney, explained that by focusing on poaching, the media "completely ignores a bigger problem . . . habitat [loss]" (Figures 5.2 and 5.3). Freelance journalist Karl Mathiesen linked media foci to priorities of predominantly international conservation NGOs, saying, "Media focus on the poaching . . . the campaign priority for the people who are feeding a lot of the information to the media." Other interviewees expressed concern about the predominance of elephant-centric narratives, which exclude the human dimensions and implications of elephant conservation. Julian Sturgeon of Resource Africa said, "Human and elephant interaction has been relegated . . . the narrative seems to completely ignore the principal of sustainable utilization." The president of the African Wildlife Foundation (AWF), Kaddu Sebunya, echoed this, saying, "Elephants are so important for the livelihoods of people, that has not been explained."

Interviewees suggested that frames using straightforward narratives result from journalists being increasingly time-poor and consequently reliant on stories and sources available electronically on short notice. Timing is everything in getting views to print. TRAFFIC's Richard Thomas explained, "If it's [an enquiry] in reaction to a press release . . . it would need a response . . . probably within the hour."

Desk-based journalism poses challenges for achieving greater representation in reporting. Thomas observed that "because of the pressures on journalists, press releases are often used as a copy and paste," and Mathiesen acknowledged that owing to an expectation to produce three or four stories a day, "it's very

Figure 5.2 "Wildlife Waste" – confiscated raw ivory on display at the Los Angeles Zoo.

Source: Photograph by Kate Ochsman, published with permission.

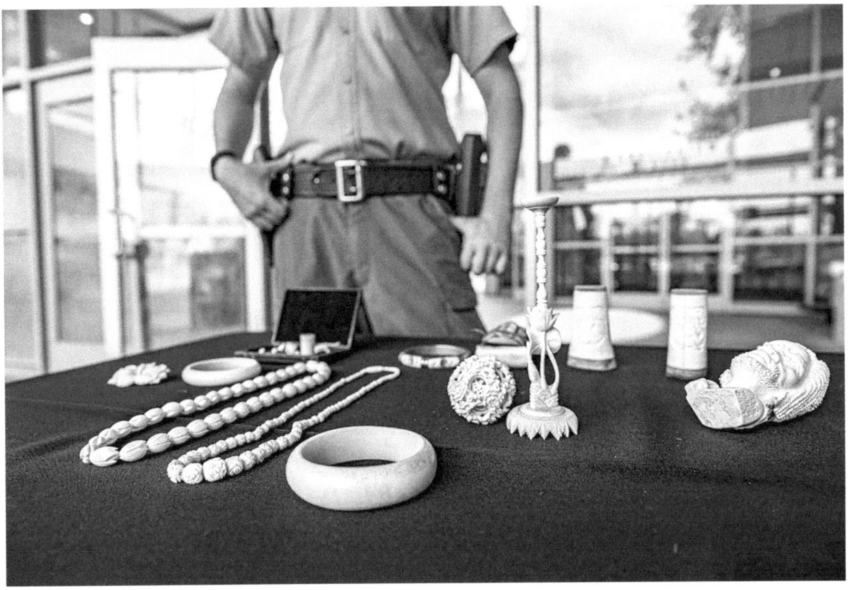

Figure 5.3 "Trinkets of Death" – confiscated worked ivory products on display at the Los Angeles Zoo.

Source: Photograph by Kate Ochsman, published with permission.

easy to just re-report press releases from the conservation movement because it . . . fits with our readers' value system."

However, an anonymous Zimbabwean journalist described having up to seven days to complete a story, including 2–3 days in the field. Content analyses demonstrated that African reporting is more likely to include socially situated discourse frames and higher numbers of African sources. It is reasonable to suggest that maintaining investigative reporting practices may allow for voices which are not so readily connected via electronic means to be heard.

Building source–journalist relationships

Source–journalist relationships are time-consuming and involve sustained participation, commitment of resources, and near-constant electronic dialogue. Sebunya outlined AWF's approach to communications, saying, "We've devolved a communications team . . . looked at different markets . . . have spokespeople . . . use social media." The engagement needed to uphold these relationships was echoed by Endangered Wildlife Trust founder Clive Walker, who said that "responding immediately to what [journalists] want" was key. Interviewees said communities they work with, or are part of, lack the time, resources, or connectivity to maintain relationships with media professionals in this way.

Barriers to diversity

On appreciating the value of media exposure, Sebunya remarked, "Within our team, yes, but we don't see that a lot within other NGOs or other communities." Stiles said, "It's too risky for any high-profile person to. . . [speak up for sustainable-use] right now. The press would probably crucify them." Community conservancy representative Lamson Malukele explained, "Not all communities understand . . . their role in terms of how conservation should look like. . .[H]istory made it a point that we look at conservation as an industry that does not belong to the black community."

Interviewees also cited a lack of practical understanding that prevents journalists from connecting with African sources. UK journalist Mathiesen recalled difficulties in contacting African governments, saying, "I didn't know how to address their officials properly . . . what the correct approach was." Cooney of IUCN SULi asked, "How do you make contact with journalists . . . write your things . . . time things so that they get picked up?"

Media and conservation policy

Interviewees perceived direct links between media coverage and policy decisions around elephant conservation. Sebunya describes AWF's strategy for communications: "If our message is targeted to politicians, policy makers, our leaders, we look at the national media houses . . . for Africa corporate . . . business media houses." At TRAFFIC, Thomas explains:

> We do a sort of quick and dirty analysis of where can we get maximum influence . . . If we're going to get published in a British tabloid . . . that's not likely to be read by the minister in an Asian country that we want to see it.

Asia is traditionally the world's biggest continental market for ivory.

Cooney says the predominance of an "action-oriented, dramatic media narrative drives a certain crisis mode of decision-making and policy making that looks at simple solutions." Solutions that tend to be prohibitive or exclusionary, such as banning sales or establishing protected areas, further drive a wedge between elephants and local communities. A second anonymous interviewee summarized the media's power to change policy, stating, "It does not always take a ban from the range states to end trophy hunting. If the importing countries ban the import, it will have the same effects. Commissioners do not want to offend their constituency."

The increased globalization of communications and conservation, referred to by Barua (2014, p. 570) as "the folding of space and time," coupled with a resulting increase in desk-bound journalism, may have caused reporting – and policy making by association – to overstate the role of international actors, such as spokespersons for international NGOs and Western celebrities, and tend towards compromise and simplicity.

Discussion

Making elephants relevant

Considering the data by continental distribution alone, content analysis supports the narrative that events "close to home" receive most media attention, with frequency and longevity of coverage decreasing as geographic distance from elephant range states increases (Bendix & Liebler, 1999). However, the United Kingdom generated the most articles for a single country. Journalists are translators, making sense of geographically or culturally distant phenomena to audiences back home (Nothias, 2017; Nyamnjoh, 2017). Therefore, perceived newsworthiness, brought about by presenting issues in ways that create social proximity to geographically distant audiences, may be equally powerful determinants of media coverage.

Although African reporting more often employed "crime/law enforcement" and "community-based conservation" frames in which Africans are central to the story, the more common UK coverage relied heavily on "international aid/partnerships" and "animal welfare" frames that highlight global actors and values. For global audiences reading syndicated or online reporting originating in the United Kingdom, this process of creating relevance actively contributes to framing elephant conservation in a way that disproportionately focuses attention on the role of external expertise and aid. At the same time, it casts doubt on Africans' own capacity to conserve their wildlife. This may broaden the divide between local people and wildlife by directing resources towards international organizations and protectionist solutions and by entrenching African beliefs that conservation is for "foreigners" (Franks, 2010; Nothias, 2017).

Overall, socially situated dimensions of the elephant crisis, including human–elephant conflict and habitat loss, were noticeably lacking in UK and African coverage. UK reporting was more likely to include "human–elephant conflict," potentially due to the shock value and novelty of this trope for international audiences (Barua, 2010). "Information," "crime/law enforcement," and "ivory demand" narratives also dominated. Conservation NGOs compete for public attention and money in a global market, and the simplicity of the poaching narrative (indicated by "ivory demand" and "crime/law enforcement" frames) offers clear heroes and villains. It is also an easier campaign focus than communicating alternate narratives of increasing complexity (such as habitat loss), which may be more difficult and less lucrative (Duffy, 2010).

Data demonstrates journalists' use of the "community-based conservation" frame is, however, on the rise. This increase is analogous to emergent academic debate around the subject in the wake of protectionist conservation failures, as well as increasing internet connectivity in rural Africa, offering hope for the diversification of narratives in reporting.

Sources and voices

Official voices overwhelmingly dominate elephant conservation reporting. Together, international conservation organizations, government and national parks spokespeople, academics, CITES, and celebrities account for nearly 90% of sources. As these sources are inherently authoritative and the ivory trade is a global political priority, it is natural that worldwide politicians and scientific expertise will contribute to reporting. However, if local interest groups or residents "warrant voice" in coverage of scientific issues of societal concern, this rarely appears to be the case in elephant conservation reporting (Gergen, 1989).

Recent sea-changes in the mass media have resulted in greater reliance on desk-based journalism and "information subsidies" readily provided by connected, well-resourced public relations professionals, including international conservation organizations that are increasingly professionalized in their media communications (Duffy, 2010). This study adds to existing evidence that media-savvy, international NGOs are aware of journalists' desire for convenient, time-saving sources of information they can reproduce with minimal editing (Fenton, 2010; Wright, 2017). Interviewees recalled press releases from international NGOs being copied verbatim into reporting. Local African conservation and community organizations lack the time, media training, and connectivity to participate equally.

This study identified ways influential individuals can influence agendas. Celebrities may help sustain media attention, raise awareness, and bring deliberative democracy to the media. However, their presence may also lend unwarranted legitimacy to certain narratives and squeeze out other voices (Street, 2004; Anderson et al., 2005). The increase in UK reporting in 2016–2017 corresponds to *The Guardian*'s year of reporting on the elephant throughout 2017. It was financed by Microsoft's Paul Allen, and the Giants Club Summit, an Evgeny Lebedev initiative uniting global leaders around elephant conservation and publicized through Lebedev-owned media outlets. Evidently, the elephant has become a powerful symbol of conservation and environmental credentials that many public figures want to be seen to champion. While this level of support is undeniably vital to elephant conservation, the level of coverage generated by the prerogatives of select wealthy individuals is concerning. Their socioeconomic status allows them to dictate media coverage to a degree that may harm the democratic production and deliberation of scientific issues in the press. This study found that celebrity presence closes down discussion, amplifies certain perspectives, such as animal welfare and international aid, and neglects others, such as community-based natural resource management and human–elephant conflict.

Readers not only acquire information from the media but also decide how much importance to attach to that information and to those who provide it, based on many factors, including hierarchical placement within reporting (Tran & Ragas, 2017).

Government/Ministerial and international conservation organization sources accounted for the greatest percentage of lead sources in reporting. These sources have accrued social capital, established authoritative relationships with journalists, and their legitimacy is reinforced through lead positioning in media texts (Munnik, 2017). Considered alongside the fact that "animal welfare" is the most common lead and headline frame, readers on both continents could believe that conservation led from the top, in isolation from local people, is the dimension of elephant conservation that warrants greatest attention and investment, neglecting other valid voices and viewpoints.

Despite 43% of lead sources reflecting "African voice," local African conservation organizations and community representatives constitute just 4% of lead and second sources, and 5% of sources in the dataset. Interviewees emphasized how important it is for Africans to hear from Africans if elephant conservation is to succeed. This study made no ethnic or racial distinction in classifying African voice. It was implicit, however, in interview responses that the voices they feel are missing from public communications around elephant conservation belong to Black Africans from non-elite groups living alongside elephants in rural areas.

In addition to practical obstacles making it physically more difficult to participate, interviewees cited ideological barriers that preclude certain people, particularly local field project staff and rural Africans, from media participation. Among these ideological barriers was a sense of "not belonging" – a vestige of the colonial development of both conservation and the media industry (Duffy, 2014; Holmes, Sandbrook, & Fisher, 2017). The predominance of scientific narratives aligned in the popular mind with white, middle class masculinity, and the absence of non-elite Africans reporting about elephant conservation may reinforce the unhelpful notion that elephant conservation "does not belong to the Black community" (Malukele, personal communication, 2017).

Community-based conservation, involving local people and drawing on traditional ecological knowledge to conserve biodiversity while enhancing livelihoods, may provide the best solutions for sustainable African elephant conservation (Ruiz-Mallén & Corbera, 2013). To raise this approach higher in the conservation agenda, African leaders, global decision-makers, corporations, and investors need to hear from those who have this knowledge and whose lives and livelihoods are most closely affected by elephants, poaching, and conservation.

Conclusion

This study looked beyond visible representations of poaching and conservation in the media and contributes empirical evidence, and contextual understanding, of forces and actors that shape the creation of elephant conservation in the media. Representations differ across continents, with journalists in the United Kingdom and Africa employing different frames and sources and arranging them according to their perception of what matters to their respective home audiences.

However, the African and UK articles are similar in important ways. Statistical information, crime, and ivory demand dominate coverage to varying degrees, effectively excluding human and socioeconomic dimensions of conservation. Moreover, elite actors account for the overwhelming majority of sources, with fewer than 6% of sources categorized as African conservation organizations or community representatives.

The fast-paced nature of contemporary media production undoubtedly influences reporting, contributing to a narrowing of perspectives and the relative absence of grassroots African voices. The privileging of narratives that focus on poaching and separate elephants from people correlates with a tendency towards prohibitive legislation, increasingly militarized anti-poaching measures, and initiatives focused on conserving the species itself rather than integrating conservation with rural development initiatives that benefit people and wildlife, and may be more sustainable in the long term. Between 1970 and 1985, the global network of officially designated Protected Areas expanded by 80%. Despite commitments by the world's largest conservation organizations and governments to find alternatives to so-called "fortress" conservation, little has changed in the decades since (Tauli-Corpuz, 2018). The media's representation of elephant conservation may help perpetuate "fortress"-style conservation initiatives that separate people from wildlife. More important, perhaps, is the media's potential to disengage local Africans from conservation by emphasizing the views of elite, often external, actors. As an arena providing publics, investors, and decision-makers with ways of understanding and figures of authority, it is essential that the media provide a more diverse representation of African elephant poaching and conservation. To avert the elephant population crisis and for *Loxodonta africana* to warrant its all-African suffix, all voices connected to the issue, and the full range of stories, must be heard.

References

Abell, J., & Myers, G. (2008). Analyzing research interviews. In R. Wodak & M. Krzyzanowski (Eds.), *Qualitative discourse analysis in the social sciences* (pp. 145–159). New York: Palgrave Macmillan.

Anderson, A., Allan, S., Peterson, A., & Wilkinson, C. (2005). The framing of nanotechnologies in the British newspaper press. *Science Communication*, *27*(2), 200–220.

Barua, M. (2010). Whose issue? Representations of human-elephant conflict in Indian and international media. *Science Communication*, *32*(1), 55–75.

Barua, M. (2014). Circulating elephants: Unpacking the geographies of a cosmopolitan animal. *Transactions of the Institute of British Geographers*, *39*(1), 559–573.

Bendix, J.,& Liebler, C. M. (1999). Place, distance, and environmental news: Geographic variation in newspaper coverage of the spotted owl conflict. *Annals of the Association of American Geographers*, *89*(4), 658–676.

Bhatia, S., Athreya, V., Grenyer, R., & MacDonald, D. W. (2013). Understanding the role of representations of human-leopard conflict in Mumbai through media-content analysis. *Conservation Biology*, *27*(3), 588–594.

Blewitt, J. (2011). The media, animal conservation and environmental education. *Environmental Education Research*, *17*(6), 711–718.

Carvalho, A. (2008). Media(ted) discourse and society. *Journalism Studies*, *9*(2), 161–177.

Chase, M. J., & Landen, K. (2019). Counting elephants. In D. Pinnock & C. Bell (Eds.), *The last elephants*. London: Hardie Grant.

Chase, M. J., Schlossberg, S., Griffin, C. R., Bouché, P. J. C., Djene, S. W., Elkan, P. W., . . . Sutcliffe, R. (2016). Continent-wide survey reveals massive decline in African savannah elephants. *PeerJ*, *4*(e2354), 1–24.

Douglas-Hamilton, I. (1987). African elephants: Population trends and their causes. *Oryx*, *21*(1), 11–24.

Dowie, M. (2011). *Conservation refugees: The hundred-year conflict between global conservation and native peoples*. Cambridge, MA: MIT Press.

Duffy, R. (2010). *Nature crime: How we're getting conservation wrong*. London: Yale University Press.

Duffy, R. (2014). Interactive elephants: Nature, tourism and neoliberalism. *Annals of Tourism Research*, *1*, 88–101.

Entman, R. M. (1993). Framing: Toward clarification of a fractured paradigm. *Journal of Communication*, *43*(4), 51–58.

Fenton, N. (2010). NGOs, new media and the mainstream news: News from everywhere. In N. Fenton (Ed.), *New media, old news* (pp. 153–186). London: Sage.

Franks, S. (2010). Globalising consciousness: The end of "foreign" reporting and no more foreign secrets. *Ethical Space: The International Journal of Communication Ethics*, 7(4), 39–47.

Gamson, W. A., & Modigliani, A. (1989). A constructionist approach to mass media and public opinion. *Symbolic Interaction*, *11*(2), 161–174.

Gardner, B. (2017). Elite discourses of conservation in Tanzania. *Social Semiotics*, *27*(3), 348–358.

Gergen, K. J. (1989). Warranting voice and the elaboration of the self. In J. Shotter & K. J. Gergen (Eds.), *Inquiries in social construction: Texts of identity* (pp. 70–81). Thousand Oaks, CA: Sage.

Hansen, A. (2011). Communication, media and environment: Towards reconnecting research on the production, content and social implications of environmental communication. *The International Communication Gazette*, *73*(1–2), 7–25.

Hivon, M., Lehoux, P., Denis, J. L., & Rock, M. (2010). Marginal voices in the media coverage of controversial health interventions: How do they contribute to the public understanding of science? *Public Understanding of Science*, *19*(1), 34–51.

Holmes, G., Sandbrook, C., & Fisher, J. (2017). Understanding conservationists' perspectives on the new-conservation debate. *Conservation Biology*, *31*(2), 353–363.

Kwansah-Aidoo, K. (2001). The appeal of qualitative methods to traditional agenda-setting research: An example from West Africa. *Gazette*, *63*(6), 521–537.

Mbaria, J., & Ogada, M. (2016). *The big conservation lie*. Kenya: Lens&Pens.

McCallum, K., & Waller, L. (2013). The intervention of media power in Indigenous policy-making. *Media International Australia*, *149*(1), 139–140.

Munnik, M. (2017). A field theory perspective on journalist-source relations: A study of "new-entrants" and "authorized knowers" among Scottish Muslims. *Sociology*, *51*(1), 1–16.

Nisbet, M. C., Brossard, D., & Kroepsch, A. (2003). Framing science: The stem cell controversy in an age of press/politics. *International Journal of Press/Politics*, *8*(2), 36–70.

Nothias, T. (2017). Mediating the distant other for the distant audience: How do Western correspondents in East and Southern Africa perceive their audience? In M. Bunce, S. Franks, & C. Paterson (Eds.), *Africa's media image in the 21st century: From the "heart of darkness" to "Africa rising"* (pp. 73–82). New York: Routledge.

Nyamnjoh, F. B. (2017). Reporting and writing in a world of unequal encounters. In M. Bunce, S. Franks, & C. Paterson (Eds.), *Africa's media image in the 21st century: From the "heart of darkness" to "Africa rising"* (pp. 33–37). New York: Routledge.

Riffe, D., Lacy, S., & Fico, F. (2005). *Analyzing media messages*. Mahwah, NJ: Lawrence Erlbaum Associates.

Rubado, M. E., & Jennings, J. T. (2019). Political consequences of the endangered local watchdog: Newspaper decline and mayoral elections in the United States. *Urban Affairs Review, 56*(5), 1327–1356.

Ruiz-Mallén, I., & Corbera, E. (2013). Community-based conservation and traditional ecological knowledge: Implications for social-ecological resilience. *Ecology and Society, 18*(4), 1–12.

Shaffer L. J., Khadka K. K., Van Den Hoek, J., & Naithani K. J. (2019). Human-elephant conflict: A review of current management strategies and future directions. *Frontiers in Ecology and Evolution*. 6. 235. doi: 10.3389/fevo.2018.00235

Shoemaker, P. J., & Reese, S. D. (1996). *Mediating the message: Theories of influence on mass media content* (2nd ed.). New York: Longman.

Soroka, S., Lawlor, A., Farnsworth, S., & Young, L. (2012[2013]). Mass media and policymaking. In E. Araral, S. Fritzen, M. Howlett, M. Ramesh, & X. Wu (Eds.), *Routledge handbook of public policy* (pp. 204–2014). Oxford, UK: Routledge.

Stiles, D. (2004). The ivory trade and elephant conservation. *Environmental Conservation, 31*(4), 309–321.

Street, J. (2004). Celebrity politicians: Popular culture and political representation. *British Journal of Politics and International Relations, 6*(4), 435–452.

Tagbo, E. (2010). Media coverage of climate change in Africa: A case study of Nigeria and South Africa. Fellowship, University of Oxford Reuters Institute for the Study of Journalism. Retrieved from https://reutersinstitute.politics.ox.ac.uk/sites/default/files/Media%20Coverage%20of%20Climate%20Change%20in%20Africa%20A%20Case%20Study%20of%20Nigeria%20and%20South%20Africa.pdf

Tauli-Corpuz, V. (2018). Statement of Ms. Victoria Tauli-Corpuz special rapporteur on the rights of Indigenous peoples to the Human Rights Council 39th session, Geneva. Retrieved from www.ohchr.org/en/NewsEvents/Pages/DisplayNews.aspx?NewsID=23595&LangID=E

Tran, H. L., & Ragas, M. W. (2017). Peer perceptions of media elites and hierarchical differentiation among financial journalists. *Journalism & Mass Communication Quarterly, 94*(1), 1–20.

Tyrrell, M., & Clark, D. (2014). What happened to climate change? CITES and the reconfiguration of polar bear conservation discourse. *Global Environmental Change, 24*, 363–372.

van Aarde, R. J., Jackson, T. P., & Ferreira, S. M. (2006). Conservation science and elephant management in Southern Africa. *South African Journal of Science, 102*(9–10), 385–388.

Vicente, P. N. (2017). Foreign correspondents in sub-Saharan Africa: Their socio-demographics and professional culture. In M. Bunce, S. Franks, & C. Paterson (Eds.), *Africa's media image in the 21st century: From the "heart of darkness" to "Africa rising"* (pp. 86–96). New York: Routledge.

Wright, K. (2017). It was a "simple," "positive" story of African self-help (manufactured for a Kenyan NGO by advertising multinationals). In M. Bunce, S. Franks, & C. Paterson (Eds.), *Africa's media image in the 21st century: From the "heart of darkness" to "Africa rising"* (pp. 147–157). New York: Routledge.

Part II

Images of endangered species

Communication and public opinion

6 Sea turtle tourism and conservation

Conflict and concord on a Hawaiian beach

Gavin Lamb

At Laniākea Beach on the North Shore of the island of O'ahu, Hawaiian green sea turtles (*Chelonia mydas*), or *honu* in Hawaiian, are easily found swimming in the nearshore areas and basking on the warm sand throughout the day. This has not always been the case. Until the 1990s, sightings of green sea turtles were unheard of on O'ahu and the other main Hawaiian Islands.

Historical records show that green sea turtles flourished across Hawai'i in the 19th and early 20th centuries and were an important food source for Native Hawaiians. By the 1970s, the nesting population had dwindled to 37 on a small atoll in the Northwestern Hawaiian Islands. The commercial fishing industry's overharvesting had decimated the population in the islands (Balazs & Chaloupka, 2004). As evidence of the species' decline became clear, the green sea turtle became protected under a 1974 state law. A few years later, in 1978, its global population became officially listed as threatened under the Endangered Species Act (ESA). The ESA listing not only forbids harming and killing sea turtles, but also makes human conduct that might disrupt the species' natural behaviors illegal.

Over the past 40 years, these conservation laws have led to a remarkably successful recovery. In 2016, a global review of the turtles' conservation status determined that it constituted a "distinct population segment" (DPS) of approximately 4,000 nesting females, increasing at a rate of 5.4% annually. In 2019, the International Union for Conservation of Nature Red List classified the DPS of Hawaiian green sea turtles as of "least concern" (Chaloupka & Pilcher, 2019). However, this success story has led to unforeseen challenges for conservation managers. With a steadily growing population in Hawai'i, a fast-growing sea turtle tourism industry also took root. Tourism operators capitalize on the public's fascination with charismatic sea turtles, promoting "ecotours" that allow tourists to experience them up-close in the wild. Images of Laniākea Beach as a sea turtle tourism destination circulate in advertisements on airplanes, at hotels, on websites, in guidebooks, and at street-corner newsstands dotting Waikiki, Hawai'i's tourist epicenter. Tourists also share "sea turtle selfies" taken on their visit to "Turtle Beach" on popular social media platforms like TripAdvisor, Facebook, and Instagram.

Today, an estimated 500,000 visitors arrive at Laniākea Beach annually. Like many other places where wildlife tourism and conservation activities intersect throughout the world, there is growing tension at Laniākea Beach as sea turtle tourism creates the potential for visitors to engage in harmful and disrespectful interactions. In response, conservation managers with limited resources collaborate with volunteer-based nonprofits to promote respectful viewing at Laniākea Beach and other popular sea turtle tourism destinations throughout Hawai'i. Mālama na Honu, or "Care for the Sea Turtles" in English, is a volunteer group that monitors the beach year-round. Every day, a few volunteers set up a perimeter of warning signs and red ropes around basking sea turtles to prevent tourists from harassing them and to encourage following "respectful" viewing guidelines (e.g., no touching, feeding, picking up, or riding them) (see Figure 6.1).

This chapter explores the nexus of sea turtle tourism and conservation activities at Laniākea Beach by tracing how different forms of offline and online media ("mediated") communication about them shape tourists' prior expectations and in situ encounters at this popular beach.

Profiting from people's interactions with charismatic wildlife through tourism, both in captivity and in the wild, has a long history (Bulbeck, 2005).

Figure 6.1 A daily scene at Laniākea Beach: tourists gather around to observe a basking Hawaiian green sea turtle while two volunteers monitor them.

Source: Image courtesy of Gavin Lamb.

Yet it is only within the past few decades that ecotourism has been promoted as a partial solution to the conservation of wildlife by claiming to establish non-consumptive and sustainable relationships among people, wildlife, and ecosystems (Adams, 2004). Around the world, wildlife conservation efforts increasingly depend on or accommodate market-based "solutions" such as ecotourism (Fletcher, 2011). This is especially true when conservation activities center on saving charismatic species whose appearance and behaviors trigger wide public appeal, such as orcas, elephants, and sea turtles (Lorimer, 2015). This increasing importance of ecotourism for conservation creates a tension for conservation managers, who are challenged simultaneously with promoting respectful appreciation for wildlife and reining in the public's over-appreciation of charismatic species when their behavior creates potentially harmful interactions for wildlife.

The ecotourism industry often touts itself as promoting eco-friendly "edutainment." Yet a growing body of ethnographic research in environmental communication suggests that the industry's claim to reconcile its drive for profit with ecological sustainability and animal welfare often falls short (Davis, 1997; Milstein, 2016; Mühlhäusler & Peace, 2001).

Statement of the problem

Over the past 20 years, Laniākea Beach has been transformed from a quiet beach mostly attracting local surfers and fishers into one of the most popular sea turtle tourism destinations in Hawai'i, if not the world, with thousands of tourists passing through on a typical day. This dramatic transformation emerged in tandem with a rebounding and thriving population of Hawaiian green sea turtles. Today at Laniākea Beach, masses of tourists park illegally along a coastal highway and cross a busy street to view them, creating a traffic jam that the local community calls "turtle traffic." Through the Hawai'i Tourism Authority (HTA), a tax-funded agency created in the 1990s, the state actively promotes tourism growth. However, it invests limited resources in sustainably managing wildlife and nature-based tourism destinations. At Laniākea Beach, this has meant almost no investment in infrastructure needed to support the rapid rise in tourism growth. The HTA also resists implementing and enforcing a mandatory ecotourism certification program that would foster a more sustainable tourism industry. "There's no reward for [HTA] to maintain our natural resources. They're graded on visitor arrivals," said Chris Barzman, the president of the Sustainable Tourism Association of Hawai'i. The nonprofit association provides voluntary ecotour certification (2019).

Against this backdrop of growing tourism and a growing sea turtle population, conservation managers devote almost no resources to monitoring and preventing harassment of the animals. This is because officials with the Sea Turtle Recovery division at the National Oceanic and Atmospheric Administration (NOAA) Fisheries, the federal agency charged with protecting marine endangered species, take the position that tourists' "low-impact" violations of the ESA (e.g., crowding or touching) do not pose a lethal threat to the species. Nor does the agency have the resources or staff needed to monitor the behavior of the huge number of tourists visiting Hawai'i's beaches every year. Sea turtle conservation managers face the challenging situation of balancing promotion of marine tourism that is respectful of sea turtles with the increasing frustration of local communities confronting unbridled tourist growth. In the case of Laniākea Beach, local residents feel "trapped at the crossroads" of unsustainable tourism growth on the one hand, and on the other, a feeling of being abandoned by the state and federal conservation agencies supposedly tasked with enforcing the protection of sea turtles (Kelly & Homcy, 2017).

"That's the conundrum," Irene Kelly, the sea turtle recovery coordinator for the Pacific Islands Region of NOAA Fisheries in Hawai'i, said. "How much money and time and effort do we spend on a perceived threat that doesn't really seem to harm the animal, but it's more of a public relations-kind of issue? We have a human tourism issue, we don't have a turtle issue so much" (2017). The public relations problem she refers to is the growing community frustration expressed in local media about tourists' disrespectful behaviors towards the sea turtles. In this context, NOAA's strategy has been to collaborate with the HTA to manage tourist expectations and behavior through marketing and messaging campaigns before they arrive at a beach and, when communities feel there is a need, to support volunteer efforts to encourage respectful viewing guidelines at popular destinations like Laniākea Beach.

To examine the problematic nexus of sea turtle conservation and tourism, this chapter examines the role of "mediated communication" in shaping people's interactions with charismatic wildlife, whether to admire, protect, or exploit animals. First, it examines tourism media about sea turtles to determine how they enable, or "mediate," how tourists learn about and behave around sea turtles. The chapter then explores whether these media reports about sea turtles are "invariably propelled/influenced and manipulated by competing agencies and interests, which in turn have very different degrees of power and very different communicative resources at their disposal" (Hansen, 2011, p. 21). It focuses on how conservation managers, in collaboration with volunteer-based conservation groups, seek to influence "turtle tourism" communication to promote lawful and ethical encounters along Hawai'i's beaches.

Methods

This study involved a 2½ year ethnography of the volunteer-based sea turtle conservation nonprofit group Mālama na Honu. The data includes interviews with a range of stakeholders in sea turtle tourism and conservation, as well as photos, field notes, and video-recordings of interactions among tourists, volunteers, and sea turtles.

The following sections explore the discourse cycles of sea turtle tourism and conservation that converge at Laniākea Beach (Scollon & Scollon, 2004). They examine how sea turtle tourism media circulate across offline and online circuits (Salazar, 2010) to inform tourists' prior expectations, in situ actions, and post hoc stories about sea turtles (Urry, 2002; see also Sowards, 2012). For example, tourists arrive in search of the sea turtles they have already seen, heard, or read about, then track down and capture similar representations to share online as "proof of having seen, experienced, and encountered the place for themselves" (Thurlow & Jaworski, 2014, p. 463). These communication sites, or circuits, include: 1) the mediatized representations of commercial sea turtle tourism advertising found in brochures, magazines, and tour websites; 2) the mediated actions of tourists' in situ behavior around people and sea turtles on the beach; and 3) the remediated practices involved when tourists refashion digital photos and videos to share on social-networking platforms.

To unpack the forms of collaboration and conflict that emerge, the chapter examines how a discourse of sea turtle tourism circulated across the three tourism circuits and then asks how conservation managers, in collaboration with community activists, intervened in these circuits to influence and challenge this discourse.

Sea turtle tourism communication

Circuit 1: mediatized representations

Mediatization refers to media that involves commodified or commercialized practices of representation. In the context of wildlife tourism, profit-driven stakeholders in the industry strategically produce mediatized representations to shape tourists' expectations and desires to experience particular places and wildlife (in this case, sea turtles at Laniākea Beach). Before ever arriving at the beach, tourists from diverse cultural and linguistic backgrounds must first learn that there is such a place so they may develop the motivation to spend money to get there. Many visitors to Hawai'i learn about Laniākea Beach through a robust commercial industry that advertises "turtle tours" in print and digital media. The HTA also promotes this commercial network of "ecotourism" media.

Images of tourist encounters with sea turtles published in the multilingual "turtle tour" brochures that are readily found in hotel lobbies, shopping centers, and brochure stands along the busy streets of Waikiki and featured on tour operators' websites, offer customers a "hub of opportunity, adventures, and experiences that are ranked highly. This is one stop shopping for the best in Hawaii" with "no tourist traps allowed" (Adventure Tours HI, 2019). One Japanese advertisement in a free print magazine at a Waikiki newsstand shows a string of pictures of tourists with turtles and reads: "Get up close and intimate with sea turtles! And snorkel with them in a fun, all-inclusive package!" (Outdoor Hawaii, 2019).

Well before tourists arrive at Laniākea Beach, these commercial representations tap into and cultivate their desires for "authentic" encounters with sea turtles in the wild. And even before tourists see such media messages, their interest has been triggered by global ecotourism promoting the natural desire to personally experience the spectacular wildlife portrayed in advertising, nature documentaries, and Hollywood films. A "compulsion to corporeal proximity" (Urry, 2002, p. 154) may be an essential ingredient in expectations for "authentic" encounters. In other words, swimming with, touching, feeding, and otherwise being within reach of sea turtles is the key here. Authentic nature in the ecotourist experience, as scholars of ecotourism point out (e.g., West & Carrier, 2004), turns on a Western ecocultural framework that values wild nature as a place defined by the absence of humans. This idea of pristine nature empty of people finds expression in the explicit imagery and text in ecotour appeals to potential clients, describing Laniākea Beach as a journey to a "secret location" far from the overcrowded "tourist traps" of Waikiki.

Circuit 2: mediated actions

The notion of "mediated action" (see Norris & Jones, 2005) draws attention to the different types of tools, whether physical, symbolic, or psychological, that make it easier or harder for a social actor to carry out particular actions and activities. Depending on the actions taken – like touching a sea turtle or posting a selfie to Instagram – mediational tools may include language, past experience, media representations, objects, technologies, and both the human-made and natural environment in which actions take place. In this case, a key symbolic tool that mediates (enables) certain tourists' actions at Laniākea Beach are commercial, mediatized representations. These media messages shape tourists' anticipations about the kinds of actions they can expect to carry out at Laniākea Beach. They expect not just to see sea turtles, but to touch, feed, swim with, or ride them. In sum, examining tourists' real-time mediated actions with sea turtles reveals how mediatized representations produced by the tourism industry enable and constrain (i.e., mediate) the kinds of actions that tourists actually carry out and deem necessary to properly "consume" Laniākea Beach.

These media representations prepare visitors to experience a wide range of activities, including snorkeling with sea turtles. When not resting on the beach, sea turtles are often foraging on the green algae growing on rocks along the shoreline, bobbing their heads up and down as they scrape *limu*, or seaweed, off the rocks with their serrated beaks. Tourists interviewed for this study made it clear that their expectations to swim with and touch sea turtles came from spectacular imagery of such close encounters advertised in tour brochures, on websites, and increasingly on social media. However, when they actually arrive, they learn how to behave around sea turtles from watching other tourists. When preparing to swim with sea turtles, they often take a moment to assess the situation. They observe how others enter the ocean, where and how far offshore they need to go to get close to the turtles, and they ask questions such as: "How many sea turtles are there?" For some tourists unfamiliar with the ocean, snorkeling is a daunting activity, and watching others do it first can alleviate concerns about entering this unknown space.

But perhaps the most anticipated action at Laniākea Beach is to get a "sea turtle selfie." It is not enough to merely see sea turtles in the wild. They need to be seen to have seen one.

Circuit 3: remediated practices

A focus on remediated practices explores the increasingly popular set of digital communicative choices that enable tourists to capture their brief, on-the-ground interactions with sea turtles and recycle or "remediate" them on online social media platforms. Remediation is "a defining characteristic of the new digital media" (Bolter & Grusin, 1999, p. 45) and involves refashioning an old representation (such as a selfie on a smartphone) into a new medium (e.g., on one's Facebook or Instagram account). Online remediation has the potential to reach a global audience with one's digital content through the communicative tools of "viral" sharing networks (e.g., hashtags, geolocations, and followers). Notably, Instagram has become one of the most popular social media platforms as a photo- and video-sharing social-networking service. Around 1 billion people use Instagram globally, with about 140 million active Instagram users in the United States (Clement, 2019), an indication of the pervasiveness of sharing digital photography on social networks.

Instagram posts of sea turtle selfies enable tourists to capture their fleeting interactions and quickly recycle them into a more enduring and visible digital medium. Indeed, this study's interviews with hundreds of tourists over the course of two years at this beach revealed that many had learned about it as a sea turtle tourism destination through social media sites, rather than through more traditional channels of commercial advertising. The popularity of sea turtle selfies reflects how charismatic wildlife are increasingly enlisted as a resource for constructing online narratives of "self," contributing to people's digital identity projects of "virtual self-branding" (Georgakopoulou, 2016).

Sea turtle conservation communication

This chapter also illustrates how wildlife conservation managers collaborate with community activists to influence mediatized, mediated, and remediated tourism communication. In doing so, they hope to promote more ethical human–sea turtle interactions at Laniākea Beach.

Circuit 1: mediatized representations

Wildlife conservation managers, faced with the proliferation of mediatized representations of sea turtle tourism, have sought to intervene in commercial media streams to influence tour operators to promote more "respectful" viewing guidelines. Since the state of Hawai'i resists regulating ecotourism, conservation managers often bypass state agencies and communicate directly with the tourism industry. "So what we're trying to do now," NOAA's Irene Kelly says, "is target more the tourist industry itself, like the tour operators to get them information because the state doesn't mandate that they have any training" (2017). For example, NOAA has convinced some English- and Japanese-language tour operators to include lawful and/or respectful wildlife viewing guidelines in their advertising materials (such as, "please maintain a respectful viewing distance of 10ft/3m from sea turtles"). NOAA also uses public service announcements, squeezed between tourism advertising on in-bound flights to Hawai'i, to circulate wildlife viewing guidelines to visitors.

The HTA may have the potential to create a more cohesive communication network of oversight and regulation for wildlife tourism messaging, but thus far has primarily promoted tourism, with few resources put into sustainably managing ecotourism destinations. A scathing state audit of the HTA in 2018 showed that the agency violated its own media policies for promoting ecotourism and wildlife tourism destinations like Laniākea Beach on Instagram and other social media platforms. The audit report "found the HTA had posted photos of people getting close to protected species, removing rocks and seashells, and other depictions of areas throughout Hawai'i that the HTA classifies as 'do not promote.'" The audit also revealed that the HTA distributed these images to tour operators for use in their commercial advertising.

Circuit 2: mediated actions

Tourists arriving at Laniākea Beach are often surprised to encounter volunteers, or "honu guardians," placing warning signs to "look, don't touch" and to "view sea turtles respectfully," with red ropes forming a protective perimeter around basking sea turtles. By doing this, volunteers hope to establish a viewing distance of 10 feet (3 meters). These communicative tools include educational signs describing sea turtle ecology (see Figure 6.2), hand-crafted signs made by community activists displaying the names of familiar individual sea turtles, and educational brochures in various languages interspersed with Hawaiian words such as "Aloha."

These signs and brochures combine educational, legal, or threat-based messaging common in the communication materials of both grassroots activism and state-based wildlife protection efforts in the United States (Abrams, Leong, Melena, & Teel, 2019). Conservationists in Hawai'i commonly enlist the Hawaiian language to name organizations (e.g., Mālama na Honu) and give local cultural weight to their messaging campaigns (e.g., View *Honu* with *Aloha*), tying Western conservation values to Hawaiian ecocultural values like "Aloha 'Āina" (commonly translated as "love of the land"). However, one Native Hawaiian conservation manager interviewed for this project viewed this practice as problematic:

> Sprinkle a little Hawaiian on it, and then it makes it authentic, it makes it genuine, it makes it cultural. To just have a Hawaiian name but not have any kind of Hawaiian world view or principles at the genesis.
>
> (Anonymous, 2019)

Figure 6.2 "View Turtles Respectfully." A sea turtle regulation sign with English–Japanese translation designed by NOAA Fisheries and placed at Laniākea Beach.

Source: From National Oceanic and Atmospheric Administration.

Volunteers often recounted stories about "problem tourists" who flaunt viewing guidelines. The vast majority of volunteers with Mālama na Honu during this study were white, monolingual English speakers from the US mainland who had moved to Hawai'i as military families or to retire. Cultural differences were regarded as one of the challenges volunteers faced in their efforts to persuade international tourists from certain countries to obey the minimum viewing distance. As one veteran volunteer said:

> Well, you know, [in] some countries it's acceptable to still kill the turtle and things like that, so why can't I get up close to it? And other times it's a spiritual thing and [for] Japanese and Chinese, it's like a spirit animal . . . It's a religious experience to see that. Where me as an American, I'm just like, that's cool, it's a wild turtle! You know, I don't have those connections.
>
> (Anonymous, 2017)

Volunteers encourage tourists to view sea turtles with respect and from a distance, but legally enforcing viewing guidelines to prevent harassment is a completely different matter. Touching sea turtles is "not really illegal because the ESA doesn't define it as illegal," Kelly says. "We don't really have a working definition of harassment, and that is very frustrating." Green sea turtles are federally protected as a threatened species under the ESA, and any human behavior that qualifies as a "take" is a violation. That means any attempt "to harass, harm, pursue, hunt, shoot, wound, kill, trap, capture, or collect, or to attempt to engage in any such conduct" (16 U.S.C. 1532(19)) is illegal (Endangered and Threatened Wildlife and Plants Final Rule, 2016). However, when considering actions that do not involve killing, maiming, or relocating a species, enforcement guidelines are far from clear and are left to the discretion of local conservation managers (Nichols, 2006). As Kelly (2017) further explains:

> We don't like to see people getting close to [sea turtles] and touching them, but the Endangered Species Act that they're protected by doesn't actually specify anything more than harassment. It says, "don't harass" . . . What does harassment mean, right? It's a very human term. It doesn't have a distance.

Circuit 3: remediated practices

Finally, an increasing concern in this digital age is how tourists learn what to expect and how to behave from social media, in addition to the commercialized media circulated by industry operators. NOAA and Mālama na Honu have Instagram accounts that circulate positive messages about respectful interactions, but they are outnumbered by tourists' "sea turtle selfies." Indeed, tourists' wildlife selfies are increasingly blamed for contributing to an exploitative wildlife tourism industry at "ecotourism" destinations (see Daly & Luce, 2019). Although there are rarely any legal ramifications for people who post sea turtle selfies in Hawai'i, NOAA has pursued legal action for harassment as an ESA violation when a particular selfie provokes enough public outrage, especially when picked up by local news media outlets.

Discussion and conclusion

In 2016, the classification of the Hawaiian green sea turtle as a distinct population segment resulted from a petition filed by a Native Hawaiian nonprofit organization. It argued that the best available science indicated a recovered population in Hawai'i, and thus the Hawaiian green sea turtle should be classified as a DPS and delisted from the ESA (Endangered and Threatened Wildlife and Plants, 2016). Usually, designating a DPS is a conservation policy tool to protect a small population of a species when the larger population does not qualify for ESA protections. In this case, the intention of the petition was the opposite: to reclassify and delist. In their response, NOAA and the US Fish and Wildlife Service acknowledged the species was indeed recovering. However, they argued that based on the best available science, sea turtles must remain listed as "threatened" due to climate change. They cited threats associated with climate change like sea-level rise that will increasingly jeopardize their main nesting grounds in the Northern Hawaiian Islands, where 95% of the population nests. In light of the two agencies' findings, the sea turtle will probably remain classified indefinitely as a "threatened" DPS because the negative effects of anthropogenic climate change in the Pacific are likely to intensify.

Tourism development does not directly threaten the sea turtle's main nesting beaches, nor does it present a lethal threat to adults in the main Hawaiian Islands, according to findings in the DPS ruling. Therefore, when it comes to a growing tourism industry, conservation managers have opted towards a mostly laissez-faire approach to managing the shifting landscape of sea turtle tourism. This approach sits uncomfortably under the current "one-size-fits-all" style of ESA protection policies that agencies are charged with enforcing, an approach that often causes more confusion and frustration than trust in local communities. For example, some residents, and in particular Native Hawaiian cultural practitioners interviewed for this project, advocate for state and federal agencies tasked with enforcing restrictions to transition towards community-based natural resource management of sea turtles. Doing this can more effectively integrate local ecological and cultural knowledge into policy.

On the other hand, in a variety of community forums, many local residents criticize state and federal agencies for failing to step up and enforce what appear to them as obvious ESA violations by tourists. One solution has been for volunteers to fill the gap. Yet, some communities resist the volunteer organization at Laniākea Beach expanding to other popular tourist beaches. In the past, some volunteers were overly aggressive in reprimanding locals who got too close to turtles, causing ill-will among some residents who describe them more as a "paramilitary organization" than an educational outreach group. "It has become a sore point between people in my opinion," says George Balazs, a retired sea turtle research scientist with NOAA and co-founder of the original volunteer group at Laniākea Beach. "I think it ought to be dismantled. I think [Mālama na Honu] should go home and allow the community and the turtles to come to whatever equilibrium exists there" (2017).

Laniākea Beach has become the focal point of these conflicts as it transformed into one of the most popular sea turtle tourism destinations in Hawai'i. The influx of hundreds of thousands of tourists to a small rural community lays bare fissures and frictions about what the future landscape of human–sea turtle relations should look like in the state. Exploring communication at the nexus of sea turtle conservation and tourism in Hawai'i raises critical questions about the different degrees of power that conservation, tourism, and local community actors hold to influence people's experiences and interactions with threatened wildlife (Hansen, 2011). By examining discourse as a communicative tool for knowledge-making, one can see how power actually works in environmental communication. Simply put, "Discourses enable certain stories to be told" (Dryzek, 2013, p. 17), and stories are powerful tools for mediating human relationships with animals and nature (Stibbe, 2015). In other words, such discourses represent "possible worlds which are different from the actual world and tied into projects to change the world in particular directions" (Fairclough, 2003, p. 123). In Hawai'i, the dominance of the state-supported ecotourism industry increasingly drives alternative stories in its profit-driven direction, while framing itself as the solution to, rather than the cause of, problematic human–wildlife relations.

There are many possible stories in Hawai'i tying the fates of humans and sea turtles to a host of factors: "overtourism," conservation policy, local community protests, Native Hawaiian rights to conservation sovereignty, and the looming threats that climate change poses to the future well-being of both people and sea turtles. Examining how tourism and conservation actors with different degrees of power circulate and contest competing discourses at different "sites of meaning-making" (Rose, 2016; see also Hansen & Machin, 2013) can shed light on whose stories predominate. Researchers and policy makers might also explore offline–online circuits where new communicative connections can be made. These connections might, in turn, promote more knowledge-sharing, trust, and accountability across governments, local communities, nonprofits, and the tourism industry.

References

Interviews

Anonymous. (2017). *Veteran sea turtle volunteer.*

Anonymous. (2019). *Native Hawaiian conservation manager.*

Balazs, G. (2017). *Golden Honu Services of Oceania: Retired NOAA sea turtle research scientist and co-founder of the original volunteer group Show Turtles Aloha at Laniākea Beach.*

Barzman, C. (2019). *President, Sustainable Tourism Association.*

Kelly, I. (2017). *Sea turtle recovery manager with the Pacific Islands region of NOAA Fisheries.*

Publications

Abrams, K. M., Leong, K., Melena, S., & Teel, T. (2019). Encouraging safe wildlife viewing in national parks: Effects of a communication campaign on visitors' behavior. *Environmental Communication, 14*, 1–16.

Adams, W. M. (2004). *Against extinction: The story of conservation*. London: Earthscan.

Adventure Tours HI. (2019, March). About us. Retrieved from www.adventuretourshi.com/about/about-us

Balazs, G. H., & Chaloupka, M. (2004). Thirty-year recovery trend in the once depleted Hawaiian green sea turtle stock. *Biological Conservation, 117*(5), 491–498.

Bolter, J. D., & Grusin, R. (1999). *Remediation: Understanding new media*. Cambridge, MA: MIT Press.

Bulbeck, C. (2005). *Facing the wild: Ecotourism, conservation and animal encounters*. London: Earthscan.

Chaloupka, M. Y., & Pilcher, N. J. (2019). "Chelonia mydas" (Hawaiian subpopulation). *The IUCN Red List of Threatened Species*. doi:10.2305/IUCN.UK.2019-2.RLTS.T16285718A142098300.en.

Clement, J. (2019). Instagram – statistics & facts. *Statista – The Statistics Portal*. Retrieved from www.statista.com/topics/1882/instagram/

Daly, N., & Luce, K. (2019). Wildlife tourism. *National Geographic, 6*, 44–77.

Davis, S. G. (1997). *Spectacular nature: Corporate culture and the Sea World experience*. Berkeley, CA: University of California Press.

Dryzek, J. S. (2013). *The politics of the earth: Environmental discourses*. Oxford, UK: Oxford University Press.

Endangered and Threatened Wildlife and Plants: Final Rule to List Eleven Distinct Population Segments of the Green Sea Turtle (*Chelonia mydas*) as Endangered or Threatened and Revision of Current Listings Under the Endangered Species Act, 81 Fed. Reg. § 20057 (final rule May 6, 2016) to be codified at 50 C.F.R. 17, 223 and 224.

Fairclough, N. (2003). *Analysing discourse: Textual analysis for social research*. London: Routledge.

Fletcher, R. (2011). Sustaining tourism, sustaining capitalism? The tourism industry's role in global capitalist expansion. *Tourism Geographies, 13*(3), 443–461.

Georgakopoulou, A. (2016). From narrating the self to posting self(ies): A small stories approach to selfies. *Open Linguistics, 2*(1), 144.

Hansen, A. (2011). Communication, media and environment: Towards reconnecting research on the production, content and social implications of environmental communication. *International Communication Gazette, 73*(1–2), 7–25.

Hansen, A., & Machin, D. (2013). Researching visual environmental communication. *Environmental Communication, 2*, 151–168.

Kelly, I., & Homcy, J. (2017, February). Trapped in the crossroads of Honu conservation. *The State of the World's Sea Turtles* | SWOT. Retrieved from www.seaturtlestatus.org/articles/2018/12/18/trapped-in-the-crossroads-of-honu-conservation

Lorimer, J. (2015). *Wildlife in the Anthropocene: Conservation after nature*. Minneapolis, MN: University of Minnesota Press.

Milstein, T. (2016). The performer metaphor: "Mother Nature never gives us the same show twice." *Environmental Communication, 10*(2), 227–248.

Mühlhäusler, P., & Peace, A. (2001). Discourses of ecotourism: The case of Fraser Island, Queensland. *Language and Communication, 21*(4), 359–380.

Nichols, K. A. (2006). Turtles and tourism: Where the Endangered Species Act ends and community activism begins. *UCLA Journal of Environmental Law & Policy, 25*, 411.

Norris, S., & Jones, R. H. (2005). *Discourse in action: Introducing mediated discourse analysis*. London: Routledge.

Outdoor Hawaii. (2019, March). North Shore eco-tour [Print advertisement]. *Kaukau Hawaii Magazine*.

Rose, G. (2016). *Visual methodologies: An introduction to researching with visual materials*. Thousand Oaks, CA: Sage.

Salazar, N. B. (2010). *Envisioning Eden: Mobilizing imaginaries in tourism and beyond*. New York: Berghahn Books.

Scollon, R., & Scollon, S. B. K. (2004). *Nexus analysis: Discourse and the emerging internet*. London: Psychology Press.

Sowards, S. K. (2012). Expectations, experiences, and memories: Ecotourism and the possibilities for transformations. *Environmental Communication*, *6*(2), 175–192.

Stibbe, A. (2015). *Ecolinguistics: Language, ecology and the stories we live by*. London: Routledge.

Thurlow, C., & Jaworski, A. (2014). "Two hundred ninety-four": Remediation and multimodal performance in tourist placemaking. *Journal of Sociolinguistics*, *18*(4), 459–494.

Urry, J. (2002). *The tourist gaze*. London: Sage.

West, P., & Carrier, J. G. (2004). Ecotourism and authenticity: Getting away from it all? *Current Anthropology*, *45*(4), 483–498.

7 The 25 most endangered primates list

Impacts on conservation fundraising and policy

Kim E. Reuter, Russell A. Mittermeier, Christoph Schwitzer, Gráinne McCabe, Anthony B. Rylands, Leandro Jerusalinsky, William Konstant, Daphne Kerhoas, Jonah Ratsimbazafy, Karen B. Strier, Amanda D. Webber, Elizabeth A. Williamson and Jessica Wise

Lists of threatened species are often employed as media outreach tools, but their usefulness and efficacy are rarely tested. One prominent example is the "Primates in Peril: The World's 25 Most Endangered Primates" list – the "Top 25 List." It is published every two years by the Primate Specialist Group (PSG) of the International Union for Conservation of Nature's Species Survival Commission (IUCN SSC), in collaboration with the International Primatological Society (IPS). Eighty-eight primates have appeared in the list's 10 editions since it began in 2000 (Mittermeier, Konstant, & Rylands, 2000; Mittermeier et al., 2009, 2012; Schwitzer et al., 2014) (see Figure 7.1). The Top 25 List's purpose is to draw worldwide attention to highly endangered primates. It targets not merely a supportive public, but governments, donors and funders, researchers, and conservation NGOs.

Twenty-five primates are included on each iteration of the list for a range of reasons. It is not simply a ranking of primates with the fewest individuals alive in the wild. This is because using a single standard metric of endangered status – such as the IUCN Red List – still results in a list of hundreds of species that are among those most threatened with extinction. Of the 706 recognized primate species and subspecies, 239 (34%) are classified as critically endangered or endangered (IUCN, 2012). While all species on the Top 25 List are officially classified as endangered or critically endangered, other factors are also considered, including lack of existing conservation attention and possible benefits from media exposure (Nekaris, Williams, Campera, & Nijman, 2019). It is also intended to ensure balanced geographic and taxonomic representation, namely, the inclusion of species from different parts of the world and different primate families and genera (Nekaris et al., 2019). Species on the list are grouped into the four major regions for primates recognized by the PSG: the Neotropics, Africa, Madagascar, and Asia.

Figure 7.1 The cover of the biennial 2018–2020 "Primates in Peril" report shows a Tapanuli orangutan, which was first described as a distinct species in 2017 and is critically endangered.

Source: From Global Wildlife Conservation, published with permission.

A small group of PSG members drew up the first list and wrote a report outlining its objectives. A paragraph on each of the 25 primates explained the reasons for their exceptional endangerment. The list was launched in January 2000 in Washington, DC, and received widespread coverage in a media environment already saturated with millennial news. With a cover and a four-page spread in *Time* magazine, the media advisory announcing the pending release, and the subsequent press coverage resulting from the launch, the list reached an audience greater than 116 million people by the end of the month (Conservation International, 2000).

The list's success motivated the PSG to repeat the exercise, this time consulting with primate researchers and conservationists in deciding which species to include. At a meeting during the 19th IPS Congress in Beijing, primatologists contributed information fresh from the field and helped decide the list's composition. The result was an updated list and report (labeled 2002–2004) in Johannesburg, South Africa, at the Rio+10 Earth Summit in September 2002 (Mittermeier et al., 2002). The revision of the 2000 list culminated in its official endorsement by the IPS, whose mission includes conservation of all primate species. The Top 25 List then became a joint endeavor of the SSC Primate Specialist Group, the IPS, and Conservation International until 2016, with Global Wildlife Conservation becoming the third partner from 2017 onwards. From the outset, the Margot Marsh Biodiversity Foundation – founded in 1996 and a major supporter of projects on apes, monkeys, tarsiers, and prosimians – has supported the list by funding primatologists' participation in biennial IPS congresses, where the list is decided, and by publishing it.

Since 2002, the list has been composed through debate, voting, and consensus by the world's leading primatologists with first-hand knowledge of the threats faced by primates in the wild. More than 250 experts were involved in compiling the last five iterations (Acerbi et al., 2020). As a result, hundreds of primatologists have met during the course of IPS congresses in lively sessions debating which species to include.

By the time the third edition was launched in 2005 in Antananarivo, Madagascar (Mittermeier et al., 2006), the presence of particular primates on the list was helping organizations supporting those species to raise funds and influence authorities. As an example, the continued profiling of Vietnamese primates caused the government of Vietnam to issue a special series of postage stamps depicting them and to launch the stamps at a special event during the 25th Congress of the IPS in Hanoi in 2014 (Figure 7.2). The increased attention encouraged, and even inspired, the government to pay more attention not just to the primate species but to the conservation of their other rich and highly endemic fauna.

More than 50 new funding sources for primate conservation were established in the decade following the launch of the list – roughly one every other month – and many emphasized support for the most endangered primates. For example, the Arcus Foundation and the Critical Ecosystem Partnership Fund were founded in 2000. The former is now the global leader in

Figure 7.2 A set of postage stamps issued by the government of Vietnam in 2014, featuring "Top 25 List" primates.

Source: Photograph by Jill Lucena, published with permission.

supporting conservation of great apes and gibbons, while the latter is a leading supporter of primate conservation initiatives in biodiversity hotspots (Mittermeier & Rylands, 2017). In addition, numerous sources emerged to generate support for Madagascar's threatened lemur fauna, including Italy's Nando Peretti Foundation, the Bristol Conservation and Science Foundation in the United Kingdom, and at least half a dozen French organizations. In Brazil, the Chico Mendes Institute for Biodiversity Conservation funded projects focused on threatened Atlantic Forest primates, while several new funding organizations (e.g., the Rufford Foundation, Sea World & Busch Gardens Conservation Fund, and the Mohamed bin Zayed Species Conservation Fund) gave high priority to grant applications focused on primates from the list.

Statement of the problem

Production of the Top 25 List is no small undertaking. It involves consultation meetings with hundreds of experts, production of communications materials, and extensive media relations campaigns. In a landscape of limited conservation funding, investment of resources into the Top 25 List is justified by arguments that it may stimulate interest, facilitate fundraising, and focus government attention on key species in need of policy action. While there is anecdotal evidence in support of these claims, they have not been examined in detail.

Researchers have examined whether inclusion on the list affected scientific publications, media coverage, and interest by the general public (Acerbi et al., 2020). After analyzing a database of 40 million articles from major scientific publishers, the authors found an increase in the number of research papers mentioning a species after its inclusion on the list. They also analyzed media penetration (data from Google News) and online interest (data from Google Blogs and Twitter), collecting daily data for one month before and after the official launch of the 2014–2016 list (Schwitzer et al., 2015). The results showed a short spike of interest for a few days on Google News and Twitter but no long-term effect across the course of the month. This suggests the list had a limited impact on raising interest among the general public in these social media outlets. This analysis, however, did not assess whether the list helped improve conservation work on the ground.

This chapter examines whether the list has affected funding mobilized towards species conservation efforts. In addition, it examines whether the list has affected conservation policy and governance. This study used qualitative and quantitative data collected through a survey of members of the IUCN SSC Primate Specialist Group and by analyzing historical grant archives of seven funders of primate research and conservation programming from 1993 to 2016. The chapter includes anecdotes about impact from specialists in Madagascar, the Neotropics, and Asia to improve a contextual understanding of whether threatened species lists are an effective tool to promote primate conservation.

Methods

Survey of experts

In September 2020, an English-language online survey was sent to PSG members. The voluntary, 15-minute survey asked respondents for their opinion on the impact of the list on species conservation efforts, focusing on fundraising and policy, through a series of closed- and open-ended questions. As part of the survey, experts were asked to answer Yes or No to a series of questions that were followed by open-ended requests for additional information if the answer given was "Yes." Experts were also asked what factors they felt led to a species benefiting or not benefiting from inclusion.

1) In your opinion, did inclusion of the species on the Top 25 endangered primates list help in mobilizing resources and funding towards research and conservation?
2) In your opinion, did the inclusion of the species on the Top 25 endangered primates list help in any way in terms of policy or governance actions to protect the species?
3) In your opinion, has inclusion in the Top 25 endangered primates list galvanized conservation action for the species and/or would you describe inclusion in the Top 25 List as a turning point for conservation of the species?

The 650 members were twice invited to complete the survey. Respondents were asked to reflect on the impact of the list on a specific primate taxon which they had expertise with. They had the option to give information on more than one primate (i.e., to provide multiple accounts of how the list impacted different species). In total, 113 respondents provided 200 species-specific accounts of the list's impact.

Analysis of historical grant archives

The authors compiled historical information from seven grant-giving programs to understand whether and how the list affected research and conservation funding. They compiled the archives of seven entities: 1) the American Society of Primatology; 2) the Margot Marsh Biodiversity Foundation, one of the world's largest grant-givers focusing on non-human primates; 3) the Mohamed bin Zayed Species Conservation Fund; 4) the Nacey Maggioncalda Foundation; 5) the Primate Action Fund; 6) Primate Conservation Inc.; and 7) the Primate Society of Great Britain.

These funding sources were selected from a pool of 46 donors/institutions because they could provide searchable archives and because they provide a representative cross-section of funding for primatological research and conservation programming. The Mohamed bin Zayed Species Conservation Fund, Margot Marsh Biodiversity Foundation, and Primate Action Fund explicitly consider the Top 25 List when making funding decisions.

The authors compiled data on 3,278 research and conservation grants from 1993 to 2020 for work on 405 primate taxa. Given that the impact on research funding from inclusion on the list can take several years to become apparent, this analysis focused on a subset of this dataset: grant records up to 2016 (3,047 grants) and, in some analyses, on species first included on the list between 2000 and 2014 (802 grants).

Findings

Impact of the Top 25 List on resource mobilization

Of the 113 respondents, 94% belonged to the Primate Specialist Group and 59% had been involved in putting together the list in the past.

In two-thirds of the accounts – 129 of 200 species-specific accounts – experts said they felt inclusion helped mobilize funding and other resources towards research and conservation. When asked for additional information on how the list supported fundraising efforts, 48 accounts provided information on a wide range of funding sources, often listing multiple funding streams and entities. Funding was most commonly secured from foundations and nongovernment grants (with 56 mentions of granting bodies across the 48 accounts). In fewer instances, funding came from governments (12 accounts, including from Indonesia, Brazil, China, the United Kingdom, and the United States), zoos (11 accounts), individual donors (4 accounts), the private sector (2 accounts, from two energy companies), and universities (2 accounts). Although most experts did not indicate total funding mobilized (and therefore the total amount of funding mobilized to Top 25 List species following inclusion on the list cannot be estimated), 35 accounts provided details of $3.87 million in grants mobilized for 33 Top 25 species following a species' inclusion and attributed by the expert to its appearance on the list.

Where inclusion resulted in improved fundraising success, this was broadly felt to be due to the ease with which the list conveyed a clear message about a species' urgent conservation needs. "Inclusion on the Top 25 List makes for strong grant proposals and makes it more likely for conservation-based projects to get funded," one expert wrote. Another noted that the list "made the marketing of the threat to red colobus species . . . much easier. It summarizes the risk in an easily digestible way for the general public and funders." A third expert said, "In terms of funding, the phrase 'listed as one of the Top 25. . .' was always included in proposals . . . It is a clear and easy way to communicate to non-technical stakeholders about how desperate things are."

A number of experts described how improved fundraising due to the list was crucial for subsequent conservation achievements. For example, inclusion of the Olalla's titi monkey was a "key aspect to obtain funds for the creation of a municipal protected area where this species is a conservation priority" in Bolivia, while in Madagascar the presence of several Top 25 species helped make the fundraising case for a large-scale reforestation program, one of the largest of its kind in the country.

In addition to the survey findings, analysis of grant archives showed that Top 25 species received a large portion of the awards from the seven grant-giving institutions. Of the 2,444 awards for work on individual species, 40% went to Top 25 List species, while the remaining grants were split across more than 300 other primate taxa not included on the list.

The number of grants awarded for any given species increased after it was included on the list. For the 63 species featured for the first time between 2000 and 2014, 604 out of 802 grants (75%) went to species after their first appearance. That means three-quarters of grants were awarded only after a species was included on the list. This is significantly different than would be expected if the list had no impact on funding, where one would see 50% of grants awarded before inclusion and 50% awarded after inclusion (Pearson chi-square test, DF = 1, $\chi^2 = 205.53$, $p < 0.0001$).

The analysis of archives also provided useful information on when species were likely to be awarded grants, relative to the year they were first listed. This provided insights into the "window of opportunity" for funding that experts felt could open following inclusion. Across the 63 species and including grants awarded both before and after inclusion, grants were awarded an average of 3.6 ± 0.45 years (± 95% confidence interval) after inclusion. Three-quarters of the grants for work on a species were awarded within eight years of its first inclusion.

Impact of the Top 25 List on conservation policy and governance

Just over half of the accounts – 102 of 200 – indicated that the list had helped policy and conservation programming on the ground. These 102 accounts listed tangible ways in which experts perceived the list to have helped policy and conservation programming, including:

1) development of conservation action plans and strategies by governments and conservation entities, as well as inclusion on national protection lists (33 accounts);
2) establishment of new protected areas or expansion of existing protected areas (30 accounts);
3) increased general government interest, including setting up national committees to address conservation concerns (27 accounts);
4) increased engagement with local human communities on conservation, including through ecotourism initiatives (13 accounts);
5) increased awareness and action to combat illegal wildlife trade, including the passing of new laws (nine accounts);
6) influencing development, mining, and infrastructure activities (six accounts); and
7) translocation, reintroduction, and captive breeding efforts (three accounts).

More than 72% of the time – in 92 out of 127 accounts – respondents reported that the list had positive impacts on policy and governance on-the-ground, where fundraising had also benefitted. This was significantly higher than the 12% of the cases (eight of 65 accounts) where experts saw no positive impact on fundraising but still felt that the list had improved policy and governance (Pearson chi-square test, DF = 1, $\chi^2 = 59.272$, $p < 0.0001$). In other words, when the Top 25 List was credited with helping increase fundraising success, it was also seen to have a positive impact on conservation policy and governance fronts.

In addition, in 80 out of the 102 responses where experts felt the list had resulted in positive impacts on policy and governance on the ground, they agreed that the impact helped "galvanize conservation action" or could be described as a "turning point for conservation for the species." Examples included government entities issuing decrees so that listed species would be recognized as a "symbol species of the State" or the list influencing decisions to implement new anti-hunting laws. One expert wrote, "In 2019, a technical review of the General Hunting Law was carried out [in Guatemala] and. . . *Ateles geoffroyi* (as a species on the Top 25 List) was promoted to an [e]ndangered species [resulting in] a total ban on hunting."

Also in the Neotropics, another expert wrote:

> When I started working on primates, *Ateles hybridus* [brown spider monkey] was not even on the national or international agenda for conservation.

There were very few biological studies and no confirmed populations in protected areas. Since the species was put on the list in 2004, many studies have been published, a conservation plan was drafted, conservation areas were prioritized, and land has been purchased for a protected area in the Magdalena Valley.

There were also many instances in which the list was thought to have had marginal or no positive impact on policy and governance. Fifty-three accounts noted reasons why a species did not benefit as much as expected. Eighteen accounts mentioned complex political and socioeconomic landscapes, including corruption and lack of government capacity and stakeholder willpower. Other reasons were:

1) issues around high-profile species – either being so well known prior to inclusion that the list had no impact or being overshadowed by more "charismatic" species (10 accounts);
2) low visibility of the list among stakeholders of interest (nine accounts);
3) ongoing threats that could not be addressed, such as hunting and deforestation (nine accounts);
4) lack of funding (eight accounts);
5) lack of existing conservation initiatives to readily scale up quickly after inclusion (four accounts);
6) unintended consequences of delisting a species (two accounts);
7) lack of data (one account); and
8) incongruence between the Top 25 List and other lists, including the IUCN Red List (one account) and a national species protection list (one account).

In a handful of cases, experts touched on unintended consequences of the list, both positive and negative. Several noted that conservation efforts for "flagship species" following inclusion helped other species indirectly as well. For example, one expert wrote:

> Inclusion on the Top 25 List helped validate the establishment of a Brazilian NGO, the *Instituto de Pesquisas Ecológicas*, which was created specifically to deal with the black lion tamarin's situation [of being threatened with extinction], but ultimately expanded to include programs focused on other Brazilian species.

In addition, several experts indicated that the list seemed to motivate young researchers, volunteers, and academics in their efforts. For example, inclusion of the Ollala brothers' titi monkey (*Plecturocebus ollalae*) in the 2018–2020 list – the first time a primate from Bolivia was included – saw the creation of the Bolivian Primatological Network within months. Two experts indicated, however, that the process of removing species from the list had negative unintended consequences. For example, one wrote, "When [the species] was taken off the list, there was a blowback. The good work done resulted in a perceived 'loss of importance' of the species."

Discussion and conclusions

Impact of the Top 25 List on resource mobilization

Most surveyed experts agreed that the list helped mobilize funding to support species conservation efforts, and this seems to be reflected in funding trends captured in the analysis of grant archives. The positive impact on fundraising may be because inclusion signals to donors and the general public that a problem needs addressing in a way that is simple to capture in a fundraising proposal.

The impact on fundraising may be especially true for species that lacked a high profile prior to inclusion. For example, *Lepilemur sahamalaza*, the critically endangered Sahamalaza sportive lemur endemic to Madagascar, was included in 2006 (Mittermeier et al., 2007), the same year that scientists first described the species. Inclusion was critical to secure funding for on-the-ground conservation programming. Crucially, the Bristol Zoological Society and the *Association Européenne pour l'Etude et la Conservation des Lemuriens*, a charity run by European zoos, supported these efforts. These groups led time-intensive fundraising efforts and provided technical support to on-the-ground programming. Awareness-raising efforts were so successful that it has not reappeared on the list since, with members of the PSG deciding to delist it so other species can be included and receive a boost in media attention. This highlights that being on the list, coupled with an established support network of researchers and conservation institutions able to mobilize quickly to reap the benefits of the list, can be an effective conservation strategy.

It can be difficult, however, to maintain the initial fundraising and research interest sparked by the list over time. This was the case with the purple-faced langur (*Semnopithecus vetulus*), an endangered primate in Sri Lanka. News in 2004 that the species had been listed among the most endangered non-human primates (Mittermeier et al., 2006) triggered a field survey to discover the reasons for the decline of its populations (Rudran, 2007). Over the next 12 years, a new community development program began with $160,000 from the SPEARS Foundation (Rudran, 2020). This program aimed to reduce human impacts on the langurs by establishing a locally managed conservation area and addressing human education, health care, and income-generation needs. As a result, local communities were brought into wildlife conservation dialogues, a critical step towards improving purple-faced langur outcomes. Limited funding, however, remains a problem, although the species has remained on every edition of the list since 2004. This suggests that the benefits of inclusion can be highest initially but decrease over time.

Impact of the Top 25 List on conservation policy and governance

Where the list resulted in increased resource mobilization for a species, it may also increase government engagement on conservation issues. Furthermore, in most cases where experts felt that conservation policy or governance actions occurred as a result of the list, they described these as turning points that galvanized efforts on the ground.

As one example, the critically endangered Hainan gibbon (*Nomascus hainanus*) was featured on the list five times between 2000 and 2016 (Schwitzer et al., 2017). Its inclusion in the 2006 edition stimulated unprecedented media interest and conservation action by the Chinese government. This was due to a large-scale effort that included a site visit to the gibbon's habitat by the chair of the IUCN SSC Primate Specialist Group, discussions with local authorities and experts, and a public presentation in Haikou, the capital of Hainan province, during which the primate was referred to as the "giant panda of Hainan." A subsequent communications campaign kicked off development of emergency action plans for the species, and funding was mobilized by IUCN, the Zoological Society of London, and the Arcus Foundation, alongside Chinese institutions. Interest in the species continues, with a recent development being a 2020 symposium in Haikou involving dozens of experts, led by the president of IUCN, to develop a strategy for the future of this species. These efforts have benefitted the species, which now numbers more than 30 individuals in the wild. Concurrently, the government committed itself to setting up a new national park to further protect the population, which currently lives only in the Bawangling National Nature Reserve.

Nevertheless, criticisms have been voiced about the list and its utility in influencing policy and governance. First, as captured by the survey, there were many accounts in which the list did not result in improved fundraising or conservation and policy impact, or where the impact was lower than expected. The difficult political and institutional contexts in range countries was the most commonly cited reason why it failed to have the expected impact. For example, one expert wrote:

> Unfortunately, the eastern [part of the Democratic Republic of the Congo] is in such a constant state of turmoil, that it is very difficult to make an impact on [a species'] conservation status unless said conservation is intense, *in situ* support. Elaborating and implementing important policy initiatives in the region is [more] challenging than in most other areas.

Complex social, economic, and political landscapes are not unique to primate conservation. For this reason, the list was designed only to call attention to the need for conservation programming of particular species and not to address social and general conservation issues within a particular region.

Other challenges noted by experts included ongoing threats, such as deforestation and hunting, and a lack of conservation initiatives and supporting organizations ready to launch communications campaigns around a species, following inclusion. As one expert noted:

> If there are not enough people working on a taxon, NGO interest, political will, and government lobbying, then this list is only so effective. Also,

money and effort tend to skew towards charismatic taxa regardless of whether or not they are on this list.

Interestingly, a number of experts felt that the visibility of the list, especially recent editions, was not high enough to influence stakeholder action. Perhaps this is because the nature of the world's media has shifted drastically over the lifetime of the list. For example, when the list first launched, only 52% of American adults were using the internet – compared to 90% of American adults in 2019 (Pew Research Center, 2020). The list was designed to make distant places and animals seem relevant to Western audiences and was a helpful tool for journalists to report about primate conservation issues to audiences that received much of their news through print media. This print media in turn legitimized the list's importance and conveyed urgency for action to key stakeholders in the private sector, governments, and grant-giving bodies.

Nowadays, more than two-thirds of Americans get at least some of their news from social media, even as public trust in science-related information on social media remains low (Clarke, Reuter, LaFleur, & Schaefer, 2019). In this busy information landscape and fast-moving news cycle, media tools such as the Top 25 List face increasing difficulty in securing the kind of coverage that they leveraged in the past. There have been cases of the list being promoted on social media. For example, Leonardo DiCaprio, a well-known actor and environmental activist, posted about the most recent list (Schwitzer et al., 2019) on his Instagram account. Nevertheless, to capitalize on the new social media landscape, the list would likely have to change not only how it is communicated, but the speed with which it is produced and iterated. It took two years for publication of a few editions. This slow production cycle could inadvertently undermine the sense of urgency that the list tries to convey. In the future, the list could use social media to announce list selection closer to "real time" (such as at the conclusion of IPS congresses) and then launch the list again when supporting documents are ready for release.

The importance of collaboration across the scientific and conservation community in ensuring the list's impact is clear, given that it is compiled through a consultative process. Where disagreements among researchers and those in the conservation community have been unresolved, they have in some cases undermined the list's potential desired impact. For example, ring-tailed lemurs (*Lemur catta*) were included in the 2016–2018 edition following worrying signs of rapid population declines due to drought (Vyawahere, 2020), habitat degradation, and the pet trade (Gould & Sather, 2016; LaFleur, Clarke, Reuter, & Schaeffer, 2017). Ring-tailed lemurs, a flagship species for the Madagascar tourism industry, have a large range in the country's south. This area historically receives less conservation funding than elsewhere in the country and overlaps geographically with other endemic species affected by similar threats. PSG members felt the ring-tailed lemur's inclusion would spur a new generation of researchers to study it in the wild, instigate government action and conservation funding, and perhaps lead to productive discussions on responsible tourism. However, the public debate on whether ring-tailed lemur populations were decreasing fast enough to merit conservation attention (Murphy, Ferguson, & Gardner, 2017), and media coverage of these disagreements (Reed, 2017), was felt by some to have affected how well the list could be used to mobilize resources for conservation. Some experts felt that the public nature of the debate initially made it harder to raise funds and advocate for ring-tailed lemur conservation than before. Over the long term, however, funding was mobilized to undertake population surveys so that its status in the wild could be resolved in more detail.

The case of the ring-tailed lemur echoes a regular criticism of the list – that it does not accurately represent the world's most threatened primates (Nekaris et al., 2019). This is because the list is not limited to species considered the most threatened based solely on something like their IUCN Red List status, population size, or whether they have a restricted range. The bind in this case – i.e., restricting the list to those primates that are truly the most endangered – is that it would be difficult, and controversial no doubt, to assess. In addition, if more limited criteria were used to compile the list, it would risk becoming confined to a familiar set of species and stagnate. That action would defeat its purpose of acting as a call to arms for highly endangered species worldwide and the dynamic forces that threaten them. In attempts to be more inclusive over the years, a number of red colobus monkeys, slow lorises, and tarsiers have been placed on the list as representatives of their genera to highlight the serious and endangered status of entire groups of species. In the Neotropics, for example, the Central American spider monkey (*Ateles geoffroyi*) was listed to highlight that two, probably three, of its six subspecies are critically endangered. The listing of *A. geoffroyi* also highlighted extinction risks across Mexico and Central America because the species is found in so many countries in the region.

Given that species' inclusion has historically been decided at in-person events, where different groups of scientists "make the case" to include their preferred species, some have argued that the list depends too much on whether species are loudly championed at the selection events, and that a species' media affinity biases the list. These criticisms take on new meaning in a post-COVID world where in-person events of hundreds of experts may not be permitted. This presents an opportunity for future Top 25 Lists to be compiled using online tools that could expand inclusion in the decision-making process.

The Top 25 List is the only media communications tool of its kind that highlights the plight of primates across the world. Within that context and looking back on its evolution, inclusion helped mobilize significant financial resources for dozens of threatened species and has led to policy and governance changes in national hunting laws, government national action and species protection plans, and expansion or creation of protected areas. Given that not all listed species benefit equally, it is clear that the list alone is not a "silver bullet" for conservation, and could never be, considering the enormous task of conserving natural areas across primate range countries. Nevertheless, as a communications tool, it has had significant impacts on donor and government awareness of, and support for, primate conservation worldwide. Looking forward, the authors of this study – several of whom are leaders in compiling the Top 25 List – will consider the results of this study to ensure it has even more impact in the future, using the many new channels that social media provides.

Acknowledgments

This chapter draws on research previously published in Acerbi, A., Kerhoas, D., Webber, A. D., McCabe, G., Mittermeier, R. A., and Schwitzer, C. (2020). The impact of the "world's 25 most endangered primates" list on scientific publications and media. *Journal for Nature Conservation, 54*. doi:10.1016/j.jnc.2020.125794.

The authors thank members of the Primate Specialist Group for information that formed the basis of this chapter and the grant agencies that shared their archives. Global Wildlife Conservation's Primate Program provided support for this study.

References

Acerbi, A., Kerhoas, D., Webber, A. D., McCabe, G., Mittermeier, R. A., & Schwitzer, C. (2020). The impact of the "world's 25 most endangered primates" list on scientific publications and media. *Journal for Nature Conservation, 54*, 125794.

Clarke, T. A., Reuter, K. E., LaFleur, M., & Schaefer, M. S. (2019). A viral video and pet lemurs on Twitter. *PLoS One, 14*(1), e0208577.

Conservation International. (2000). Top 25 most endangered primates launch: Report. Washington, DC: Conservation International.

Gould, L., & Sauther, M. L. (2016). Going, going, gone . . . Is the iconic ring-tailed lemur (*Lemur catta*) headed for imminent extirpation? *Primate Conservation, 30*, 89–101.

International Union for Conservation of Nature. (2012). *IUCN Red List categories and criteria: Version 3.1* (2nd ed.). Gland, Switzerland, and Cambridge: IUCN.

LaFleur, T. M., Clarke, T. A., Reuter, K. E., & Schaeffer, T. (2017). Rapid decrease in populations of wild ring-tailed lemurs (*Lemur catta*) in Madagascar. *Folia Primatologica, 87*(5), 320–330.

Mittermeier, R. A., Konstant, W. R., & Rylands, A. B. (2000). The world's top 25 most endangered primates. *Neotropical Primates, 8*(1), 49.

Mittermeier, R. A., Konstant, W. R., Rylands, A. B., Butynski, T. M., Eudey, A. A., Ganzhorn, J. U., & Kormos, R. (2002). The world's top 25 most endangered primates – 2002. *Neotropical Primates, 10*(3), 128–131.

Mittermeier, R. A., Ratsimbazafy, J., Rylands, A. B., Williamson, E. A., Oates, J. F., Mbora, D., . . . Aguiar, J. M. (2007). Primates in peril: The world's 25 most endangered primates, 2006–2008. *Primate Conservation, 22*, 1–40.

Mittermeier, R. A., & Rylands, A. B. (2017). Biodiversity hotspots. In T. E. Lacher (Ed.), *Encyclopedia of the Anthropocene. Volume 3: Biodiversity* (pp. 67–75). New York: Elsevier.

Mittermeier, R. A., Rylands, A. B., Schwitzer, C., Taylor, L. A., Chiozza, F., & Williamson, E. A. (Eds.). (2012). *Primates in peril: The world's 25 most endangered primates 2010–2012*. Arlington, VA: IUCN, SSC Primate Specialist Group), International Primatological Society, and Conservation International.

Mittermeier, R. A., Valladares-Pádua, C., Rylands, A. B., Eudey, A. A., Butynski, T. M., Ganzhorn, J. U., . . . Walker, S. (2006). Primates in peril: The world's 25 most endangered primates 2004–2006. *Primate Conservation, 20*, 1–28.

Mittermeier, R. A., Wallis, J., Rylands, A. B., Ganzhorn, J. U., Oates, J. F., Williamson, E. A., . . . Schwitzer, C. (Eds.). (2009). *Primates in peril: The world's 25 most endangered primates 2008–2010*. Arlington, VA: IUCN, SSC Primate Specialist Group, International Primatological Society, and Conservation International.

Murphy, A. J., Ferguson, B., & Gardner, C. J. (2017). Recent estimates of ring-tailed lemur (*Lemur catta*) population declines are methodologically flawed and misleading. *International Journal of Primatology, 38*, 623–628.

Nekaris, K. A. I., Williams, E., Campera, M., & Nijman, V. (2019). The world's 25 most endangered primate list: Contribution or detraction? *Primate Eye, 127*, 10–11.

Pew Research Center. (2020). *Internet/broadband fact sheet*. Retrieved from www.pewresearch.org/internet/fact-sheet/internet-broadband/

Reed, M. (2017, June 19). Ring-tailed lemurs down by 95 percent? Maybe not. *Mongabay*. Retrieved from: https://news.mongabay.com/2017/06/ring-tailed-lemurs-down-by-95-percent-maybe-not/

Rudran, R. (2007). A survey of Sri Lanka's endangered and endemic western purple-faced langur (*Trachypithecus vetulus nestor*). *Primate Conservation, 22*, 139–144.

Rudran, R. (2020). An ethnoprimatological approach to conserving Sri Lanka's critically endangered Western purple-faced langur. *Primate Conservation, 34*, 227–233.

Schwitzer, C., Mittermeier, R. A., Rylands, A. B., Chiozza, F., Williamson, E. A., Byler, D., . . . McCabe, G. (Eds.). (2019). *Primates in peril: The world's 25 most endangered primates 2018–2020*. Washington, DC: IUCN SSC Primate Specialist Group, International Primatological Society, Global Wildlife Conservation, and Bristol Zoological Society.

Schwitzer, C., Mittermeier, R. A., Rylands, A. B., Chiozza, F., Williamson, E. A., Macfie, E. J., . . . Cotton, A. (Eds.). (2017). *Primates in peril: The world's 25 most endangered primates 2016–2018*. Arlington, VA: IUCN SSC Primate Specialist Group, International Primatological Society, Conservation International, and Bristol Zoological Society.

Schwitzer, C., Mittermeier, R. A., Rylands, A. B., Chiozza, F., Williamson, E. A., Wallis, J., & Cotton, A. (Eds.). (2015). *Primates in peril: The world's 25 most endangered primates 2014–2016*. Arlington, VA: IUCN SSC Primate Specialist Group, International Primatological Society, Conservation International, and Bristol Zoological Society.

Schwitzer, C., Mittermeier, R. A., Rylands, A. B., Taylor, L. A., Chiozza, F., Williamson, E. A., . . . Clark, F. E. (Eds.). (2014). *Primates in peril: The world's 25 most endangered primates 2012–2014*. Arlington, VA: IUCN SSC Primate Specialist Group, International Primatological Society, Conservation International, and Bristol Zoological Society.

Vyawahere, M. (2020, June 12). One-two punch of drought, pandemic hits Madagascar's poor and its wildlife. *Mongabay*. Retrieved from https://news.mongabay.com/2020/06/one-two-punch-of-drought-pandemic-hits-madagascars-poor-and-its-wildlife/

8 You better believe it

Rhino horns are a bloody risky business

Michael Scott Smith

The illegal wildlife trade is a major crisis for conservation and international security from both supply and demand perspectives. With an annual value estimated by INTERPOL of US$15–29 billion, the trade in wild animals for medicine, pets, ornaments, and food is the second-biggest driver of species extinction and the fourth-biggest form of transnational organized crime (UNODOC, 2016). In addition, the 2020 outbreak of the coronavirus was traced to a wildlife market in China selling potentially infected bats and pangolins, drawing attention to the consumption of wildlife as a global health issue. The illegal wildlife trade in Asia came under closer scrutiny amid US and Australian calls for China to close its wildlife markets (Mukpo, 2020).

Globally, rhinos have a total estimated population of between 20,000 and 30,000 for all species combined. From the start of the poaching crisis in 2008 through 2018, 9,642 rhinos (Save the Rhino, n.d.) were poached across Africa, with the epicenter in South Africa, home to 75% of the African rhino population (UNODOC, 2020). Between 2016 and 2018, 5.9 tons of rhino horn entered the illegal market destined for Asia, generating an annual illicit income of US$230 million. Of that total, authorities seized only one ton – half in Africa, half in Asia (UNODOC, 2020). A 2019 seizure in Vietnam involved 55 pieces of horn weighing 125.1 kg (275.6 pounds) with an illicit retail value of US$3.1 million smuggled from Africa in air cargo at Noi Bai International Airport in Hanoi.

Most rhino poachers come from rural, economically marginalized South African and Mozambiquan communities that neighbor South African game reserves and national parks such as Kruger National Park. Organized crime groups exploit these communities to recruit poachers (South Africa Department of the Environment, Forestry, and Fisheries [DEFF], 2019). During 2020, DEFF recorded 394 rhinos poached for their horn in South Africa, including 1,573 poaching incursions and 247 poached rhinos in Kruger National Park (South Africa Department of the Environment, Forestry, and Fisheries [DEFF], 2021).

The market in Vietnam has been blamed as the main driver (Milliken & Shaw, 2012), with China recently coming to the fore (UNODOC, 2020). At the height of the poaching from 2013 to 2017, the supply from Africa was 30 times larger than in the early 2000s, with poachers in South Africa taking over 1,000 rhinos a year. The daily rate has since dropped from nearly two (1.85) every 12 hours when poaching was at its peak in 2015 to slightly more than one (1.2) every day in 2020 (Save the Rhino, n.d.) but remains high.

Poaching not only threatens the survival of the species worldwide, but the way that poachers remove horns from live, wounded animals can be unspeakably cruel (Smith, 2015b) (see Figure 8.1). In South Africa's Eastern Cape, poachers often dart rhinos with an immobilization drug that leaves a degree of consciousness and feeling before hacking with machetes or axes at the animal's face to remove the horn. To cut through the rhino's skull, horn, and thick skin, poachers must go through a "massive degree of hacking and fighting," creating a "huge amount of trauma," then they just leave the animals to die (Fowlds, 2015).

Rhino horn's long history as a prized ingredient in traditional Chinese medicine underpins the reasons why Vietnam, and increasingly China, are the key markets. The surge in demand is tied to its high price and Vietnam's emergence from the tight purse strings of a developing country into a fast-growing "tiger" economy sporting a growing middle class. To prevent rhino poaching and the related transnational criminal trade, governments, intergovernmental organizations, environmental nongovernmental organizations (ENGOs), and conservation groups focus on two major goals: reduce poaching in the range nations and reduce demand in end-market nations. ENGO-sponsored reduction

Figure 8.1 Poachers brutally killed this South African rhino for its horn.

Source: Photograph by Dr. Will Fowlds, published with permission.

efforts in end-market nations such as Vietnam target consumers through two main tracks: behavior change campaigns and societal control through law and enforcement (Burgess, 2016). Smith (2017) studied 76 media releases from seven key ENGOs conducting rhino horn demand reduction campaigns in Vietnam. Smith (2017) found that the ENGOs use four frames in their media releases:

- Responsibility – (The most prevalent, it appears in 68% of the sample). Aiming to change behavior, the press releases convey that Vietnamese rhino horn consumers are responsible for driving the poaching in Africa, including the conservation threat to rhino species and cruelty to individual animals. The frame also targets wealthy middle-class consumers through outlining the reputational risk to companies via Corporate Social Responsibility.
- Voodoo Wildlife Parts – (occurring in 60% of press releases). Consumers need to know that so-called medicinal benefits of rhino horn are based on superstition, without scientific basis.
- Youth Empowerment – (58%). Young people should defy traditional social hierarchies and tell their elders to stop using rhino horn (Smith, 2017, 2018a, 2018b).
- Lip Service – (49%). The government must improve its poor record of enforcement of laws against illegal trade in rhino horn (WWF, 2016, November 18) to fulfill its Convention on International Trade in Endangered Species (CITES) obligations. At the CITES Standing Committee Meetings in 2016, Vietnam was criticized for "repeated failure to supply information requested on convictions and sentences" (Knight, 2017, p. 21). Despite commitments including two Prime Ministerial Directives in 2014 and 2016 to improve efforts to combat illegal wildlife trade, only one-third of serious wildlife crimes prosecuted resulted in a prison sentence, and none of the convicted or imprisoned individuals were major figures (Environmental Investigation Agency, 2016).

As a result of ENGO media campaigns, the poaching crisis and demand for rhino horn in Vietnam have received much media attention. Reasons for the demand are strange: They are deeply rooted in traditional culture and the "internally consistent, researched discipline of" traditional Chinese medicine (Cheung, Mazerolle, Possingham, & Biggs, 2018, p. 2). And they are equally rooted in recent and constantly evolving rhino horn use trends ungrounded in traditional Chinese medicine literature, including cancer treatment, hangover cures, and social consumption.

One way to begin to understand news coverage of such environmental issues is to examine media relations efforts of ENGO sources by analyzing the news framing of their press releases, the factors behind it, and its efficacy. News framing research is an immensely productive method of analyzing environmental communication, leading to better understanding of public and political opinion and deliberation on environmental issues (Hansen, 2011). Little attention, however, has gone to research in the source strategies area of environmental communication (Hansen, 2011). This chapter focuses on "Responsibility," the most prevalent frame employed by ENGOs operating in Vietnam to reduce rhino horn demand.

Statement of the problem and research questions

This chapter explores conservation groups' press releases on an aspect of the illegal wildlife trade in Vietnam. It has relevance from a cross-cultural environmental communication viewpoint in solving the massive and diverse impacts of illegal wildlife trade.

How and why advocacy groups frame issues are important aspects of research, especially when journalists use these groups as information sources. Gitlin (1980) illustrated this in his study of how the anti-Vietnam War movement in the United States framed itself to achieve its goals and how the news media and others framed the movement. He argued that media frames are "persistent patterns of cognition, interpretation, and presentation of selection, emphasis and exclusion by which symbol handlers routinely organize discourse, whether verbal or visual" (Gitlin, 1980, p. 7).

After the Vietnam War, at the same time that Gitlin's seminal book on news framing appeared, a newly united North and South Vietnam was rebuilding under a socialist/communist government with state-controlled civil society and media. Since 1986 with *Doi Moi* (Renovation), Vietnam has opened its doors to international NGOs and allowed local ENGOs to form (Wells-Dang, 2010; Mol, 2009). Meanwhile, the media remains subservient to the government (Pham, Gregorio, & Brockhaus, 2017). The 2021 World Press Freedom Index compiled by the France-based press rights advocacy organization Reporters sans Frontières (2021) ranked Vietnam 175th among 180 countries for press freedom.

The environmental movement has grown in Vietnam (Wells-Dang, 2010; Mol, 2009). It is plausible that the "emerging green sphere" (Yang & Calhoun, 2013, p. 274) – in this case, a small group of local and international ENGOs with offices in Vietnam – is the key driver of the environmental discourse and social concern there, as Yang and Calhoun (2013) argue it is in China.

Whether from African or Asian species, rhino horn has been prized for millennia as an ingredient in traditional Chinese medicine. The trade from Africa to supply Asian traditional medicine doctors has a long history. In the early 1800s, Asian buyers purchased horns in East African ports to send to traditional Chinese medicine markets in Asia, predominantly China (Beachey, 1967). The Javan rhino species, endemic to Vietnam, once was widespread throughout India, China, and Southeast Asia. The start of the 20th century saw colonial hunting wipe the species out in country after country (de Monestrol, 1952). The last Javan rhino in Vietnam was killed for its horn in 2010, with scarcely a murmur from the Vietnamese government.

Wildlife conservation, paradoxically, stems back to British hunters concerned about dwindling African game to hunt in the early 1900s. Later that century, concerns for the African rhino's survival amid a rampant transnational trade in horn were central to the 1975 formation of the Convention on International Trade in Endangered Species. CITES placed all rhino species in Appendices I and II and banned all international trade. China at the time clamped down on the trade, stopping medicinal demand. South Africa employed a seemingly miraculous translocation program to stem poaching, and the African rhino population came back from the brink of extinction.

Vietnam has been a signatory to CITES since 1994. Even so, it has been a key market nation responsible for driving poaching in South Africa since 2008. The Vietnamese public's interest in the horn began slowly around 2006, allegedly because of a rumor that a government official had cured his cancer with it. Then demand rapidly increased, driving new heights of poaching (Milliken, 2014). Two ancient cultural reasons underpin current drivers of consumption: First, Vietnam has a wildlife eating and drinking culture (Van Song, 2008) that places prestige on the consumption of rare wildlife (Drury, 2009). Middle-class businessmen communicate wealth, power, and status by consuming it in powdered form in a drink or gifting it to influential friends (TRAFFIC, 2013). Second, rhino horn has supposed medicinal properties and is an ingredient in traditional Chinese medicine (Milliken & Shaw, 2012; Smith, 2015a, 2017, 2018a, 2018b). It is used mainly as a non-scientifically proven treatment for minor fevers (Truong, Dang, & Hall, 2016).

As the poaching crisis worsened, ENGOs in Vietnam developed communication strategies to reduce demand (Smith, 2017). ENGO media outputs that reflect these strategies are a critical form of environmental communication to be researched. This is illustrated by Cox and Pezzullo's (2016, p. 16) definition of environmental communication as "the symbolic medium that we use in constructing environmental problems and in negotiating society's different responses to them." Entman's definition of news framing reflects how framing research can address important problems and response aspects of environmental communication:

> To frame is to *select some aspects of a perceived reality and make them more salient in a communicating text, in such a way as to promote a particular problem definition, causal interpretation, moral evaluation and/or treatment recommendation* for the item described.
>
> (1993, p. 52. emphasis in original)

Hence, news frames indicate what is at stake in an issue.

This chapter demonstrates and explains some cultural challenges and potential successes of ENGOs' use of the Responsibility frame. The prevalence of this frame over others in press releases, with slogans such as WildAid's "When the buying stops, the killing can too," reflects the importance that ENGOs place on making Vietnamese consumers more aware of their role in driving high levels of poaching in order to change their behavior. In the 2014–2016 timeframe of this study, there was a sense that the campaigns were failing since African poaching statistics continued to rise and Vietnam's enforcement record remained poor.

The illegal wildlife trade constitutes a transnational environmental conflict; hence international and local ENGOs work in a cross-cultural context in Vietnam, sometimes partnering and sharing ideas for media campaigns. Framing scholars agree that culture is a significant aspect of framing environmental communication (Anderson, 1997; Cox & Pezzullo, 2016; Lin, 2012; Jeffreys, 2016). Anderson (1997, p. 9) emphasizes the importance of "local culture in framing public understandings of environmental issues." To analyze cultural and other factors that influence the production and efficacy of the ENGO frames, it is necessary to understand the interaction among sites where frames appear – the text, the communicator, the audience, and the culture (Entman, 1993). It is also essential to understand factors that influence how journalists – or in this case ENGOs – frame issues: social norms and values; organizational pressures and constraints; pressures from interest groups; journalistic routines; and journalists' ideological or political orientations (Scheufele, 1999).

Following environmental communication work on source strategies of environmental pressure groups in Britain and Australia (Anderson, 1997; Lester, 2007), there has been research into news framing of environmental and illegal wildlife trade issues in Asia (Jeffreys, 2016; Lin, 2012; Pham et al., 2017). Key debates revolve around framing in different cultural and political settings. Studies from a social marketing theoretical framework, commissioned by ENGOs working in the illegal wildlife trade area, touch on news framing in rhino horn demand reduction campaigns (Burgess, 2016; Truong et al., 2016).

This chapter builds on the existing research but focuses on the interplay and influences among the text, the communicator, the audience, and the culture in the ENGOs' production of frames.

Hence, these research questions:

RQ1: How do ENGO press releases express the Responsibility frame?
RQ2: What cultural and other factors are relevant in production of the Responsibility frame?
RQ3: How effective is the Responsibility frame at achieving the ENGOs' goal in the eyes of experts?

Methods

There were three principal phases in the study, which identified frames that ENGOs employed in demand reduction campaigns. It identified influencing factors in the frames' production and provided insight into expert views of their efficacy. The first phase was a qualitative framing analysis of relevant press releases in English. The second was semi-structured interviews with ENGO staff involved with production of media releases, and the third involved semi-structured interviews with experts, such as Vietnamese journalists, editors, public relations practitioners, and illegal wildlife trade specialists based in Vietnam. The majority of ENGO staff interviews took place at the 2016 Hanoi International Illegal Wildlife Trade Conference, employing a snowball approach. In addition to interviews, speakers at the presentation of the ENGO TRAFFIC were recorded, providing 11 transcripts for analysis. Interviews were transcribed, coded, and analyzed using the qualitative analysis program NVivo.

The framing analysis of 76 hard copy press releases noted the selection, emphasis, and exclusion of information, along with framing devices such as words, metaphors, exemplars, catch phrases, depictions, and visual images (Gamson & Modigliani, 1989). To identify the frame that dominates a press release (dominant frames) and other frames present but expressed in lesser terms (competing frames), attention was paid to whether framing devices were emphasized by their placement in framing focal points (Tankard, 2001), such as headlines and photos.

ENGOs in the sample met three criteria: They were prominent, as defined by having attended the 2016 Hanoi International Illegal Wildlife Trade Conference; they had an office or partner with an office in Vietnam; and they produced media outputs addressing the rhino horn trade and archived on their websites.

Press releases were sourced from seven ENGOs (Table 8.1) using the archive on their websites. Seventy-six press releases matched the selection criteria. They were from ENGOs in the sample, were archived on the ENGOs' websites in

Table 8.1 Seven environmental NGOs whose archived press releases were studied using the archives on their websites. Their press releases, issued between January 2014 and December 2016, were strongly relevant to Vietnam's illegal rhino horn trade.*

Environmental NGO	*Website*	*Number of Press Releases in Sample*
Change	www.changevn.org/	10
Education for Nature Vietnam (ENV)	https://env4wildlife.org/	13
WildAct	www.wildact-vn.org/	11
WildAid	www.wildaid.org/tags/vietnam www.wildaid.org/	14
Trade Records Analysis of Flora and Fauna in Commerce (TRAFFIC)	www.traffic.org/home	14
WWF-Viet Nam	http://vietnam.panda.org/en/	10
Wildlife Conservation Society (WCS)	https://programs.wcs.org/vietnam/Wildlife-Trade/Wildlife-trade-news.aspx www.wcs.org/	4

* Table courtesy of Michael Scott Smith.

the media or news section, had strong relevance to the illegal rhino horn trade in Vietnam, and were published between January 2014 and December 2016.

Findings

The key frames identified in ENGO press releases, Responsibility, Lip Service, Voodoo Wildlife Parts, and Youth Empowerment (Smith, 2017), often appeared in conjunction with each other. The Responsibility frame appeared more often than other frames in both dominant and competing forms. It appeared in 52 of the 76 press releases, being dominant in 22 and competing in 30. It was the only frame that all the ENGOs employed in at least one press release.

The Responsibility frame can be classified as a generic "attribution of responsibility" frame: "Responsibility for the cause or solution is placed on political authorities, individuals, or groups" (Semetko & Valkenburg, 2000; Dirikx & Gelders, 2009). ENGOs used this frame to convey Vietnam's responsibility for the poaching crisis in South Africa. Representations of the consequences of illegal wildlife trade, such as African rhino poaching statistics, in the press releases tend to occur in close association with the Responsibility frame. This tendency corresponds with Dirikx and Gelders' (2009) findings that Responsibility and Consequence frames often worked together in European newspaper coverage of climate change. Because of ENGOs' tendency to lump Responsibility and Consequence frames together, this study includes Consequence-framed press releases in the Responsibility frame.

The sample ENGOs expressed the Responsibility frame through seven sub-themes: crisis, brutality, corporate social responsibility, national responsibility, spiritual responsibility, health, and family values. Framing devices that reflected the frame and its sub-themes included a range of words, phrases, and images often used in association.

Table 8.2 outlines the seven Responsibility frame sub-themes and their framing devices with examples of the devices used in the press releases. Crisis and brutality were strongly represented despite possible cultural problems discussed later in the chapter. Corporate social responsibility was a key sub-theme of the ENGO TRAFFIC's "Chi" campaign. Less common sub-themes are significant in terms of how and why ENGOs employed them to exert leverage on aspects of Vietnamese society and culture.

Press releases displayed the Responsibility frame by reporting the consequences to rhinos from illegal trade and consumption in Vietnam, such as the potential crisis for rhino species and the brutality of poaching, in association with depictions of demand in Vietnam. These press releases glaringly omitted research and statistics, which Tankard (2001) argues are focal points for frames, on consumption and arrests that prove a high level of trafficking and demand in Vietnam drives poaching in South Africa. The statistics presented are instead about poaching in South Africa. A significant finding is that apart from two press releases (WCS, 2014, November 3; WildAct, 2016, November 17), statistical evidence of trade and consumption of rhino horn in Vietnam was absent

Table 8.2 Responsibility frame sub-themes description and examples. Responsibility sub-themes refer to certain impacts in Africa and locally caused by Vietnam as the key consumer nation for rhino horn. Responsibility is the most common frame used by environmental NGOs in the country to lower the demand for rhino horn.*

Responsibility Frame Sub-themes	*Meaning of Sub-theme*	*Examples of Text or Photos from Press Releases*
Crisis: Personal responsibility for crisis	By buying rhino horn individuals could contribute to the extinction of the species (ENV, 2014, September 8)	"If the killing of rhinos is not urgently stopped, rhino populations all over the world will be pushed to extinction."
Brutality: Personal responsibility for brutality	By buying rhino horn buyers are contributing to great animal suffering (ENV, 2014, September 8)	"where they might encounter poachers or come upon the grisly scene of a recently killed rhino."
CSR: Corporate Social Responsibility	International reputation of corporations is important in Vietnam's World Trade Organization membership context (TRAFFIC, 2015, June 15)	"VCCI and Traffic . . . today signed a Memorandum of Understanding (MoU) supporting the roll out of best practices in corporate ethics and consumer behaviour change, as well as a zero-tolerance approach toward wildlife consumption, through CSR."
National responsibility	By being the world's biggest consumer nation Vietnam's reputation internationally will suffer (WildAct, 2014, October 16)	"Vietnam is the largest rhino consumption nation! I feel really embarrassing, extremely embarrassing (sic). Vietnamese seems to be too ignorance (sic)."
Spiritual responsibility	Karmic impact of using rhino horn might be negative (WildAid, 2016, September 22, photo caption)	"WildAid Vietnam organises a recent event at a Buddhist pagoda in Vietnam where dharma talks encouraged Buddhist followers to protect the rhino."
Family values: Responsibility for families	By buying rhino horn buyers are orphaning baby rhinos.	Photos of family groups or rhinos with baby rhinos.
Health: Responsibility for health	By buying rhino horn, buyers risk health problems from zoonotic diseases or poisoned horns. (WildAct, 2016, June 20)	"The doctor said they [zoonotic diseases] are diseases that can be transmitted from wild animals to humans."

* Table courtesy of Michael Scott Smith.

from the 52 that incorporated the Responsibility frame. Although the press releases omitted available research, they conveyed responsibility in the crisis sub-theme by portraying a causal theme using rhetorical structures or causal statements (Pan & Kosicki, 1993). Rhetorical structures included ENGOs, such as ENV, communicating South African rhino poaching statistics and the extinction of Vietnam's own Javan rhino species due to poaching:

> During the first eight months of 2014, 695 rhinos have been slaughtered. If the killing of rhinos is not urgently stopped, rhino populations all over the world will be pushed to extinction within the next couple of years, following in the path of Vietnam's own rhinos, lost forever in 2010.
>
> (ENV, 2014, September 8)

Examples of causal statements include: "Demand for rhino horn in Asia including Vietnam is the main driver of the rhino poaching crisis in South Africa" (TRAFFIC, 2015, June 15) and "Last month because of its role as the main destination country for trafficked rhino horn" (WWF, 2016, November 18).

This excerpt outlining the ENGO WildAct's research results about online trade is one of the only two examples of any evidence in an ENGO press release:

> Almost 2000 adverts and 3000 comments were analy[z]ed . . . 38% of all advertisements found were elephant ivory products, whereas one out of every four accounts created primarily to sell elephant ivory were also selling rhino horn.
>
> (WildAct, 2016, November 17)

Other sources of statistical evidence of Vietnam's responsibility were publicly available but missing from press releases. This included sources that provided evidence of high levels of Vietnamese nationals' involvement in smuggling rhino horn in Vietnam and internationally (see Milliken, 2014; Environmental Investigation Agency, 2016), as well as proof that rhino horn is widely bought and consumed in Vietnam (see TRAFFIC, 2013; Truong et al., 2016).

One interview with a newspaper editor who chose to not be identified suggests that rhetorical or causal statements weren't effective at convincing him that Vietnam was the key market. He disagreed with the ENGOs' view, agreeing instead with information from the government that Vietnam was a transit, not consumer, nation, saying:

> Vietnam is a trans-shipment point in Asia, and you know China is the biggest consumer of rhino horns, because traditional medicine is a very strong sector in China but in Vietnam is not a strong one.
>
> (Newspaper Editor A, personal communication, November 26, 2016)

Culture has a significant influence in the production of the Responsibility frame. Both IENGOs and local ENGOs employ locally based Vietnamese to work on campaigns. Wildlife Conservation Society policy coordinator Duong

Viet Hong explained that local practitioners are more likely than foreigners to be aware of cultural nuances, such as the Vietnamese fondness for consuming rare wildlife:

> You always need to have the local people to check to see it fits with the cultures . . . Of course, the first is the office. We are the local people. We understand the messages and then [we check with] our friends and families.
> (Duong Viet Hong, personal communication, November 18, 2016)

The influence of culture can be seen in the production of the crisis sub-theme. Due to local understanding of the Vietnamese culture of eating and drinking rare wildlife, Duong Viet Hong explained that ENGOs exclude information about rarity and crisis from their media outputs:

> I tend not to use crisis [in press releases] a lot, because it might have a counter effect. Because if it's going to be extinct, [or if] someone said it's going to be extinct [the public may feel they] need to catch it now.
> (2016, personal communication, November 18)

News articles about the trade often emphasize that rhino horn is more expensive than gold or cocaine (e.g., see Harper, 2015). This is despite wholesale prices averaging US$25,000 per kilogram (UNODOC, 2020), much less than the oft-reported US$65,000. It is important to note that no press release contained information about the high price. Duong Viet Hong said high prices connote the horn's rarity and "preciousness," so WCS avoids communicating the high price as it could motivate demand (personal communication, November 18, 2016).

On the other hand, the ENGO Change demonstrated that cultural beliefs about rarity and consuming wild animals have been slowly changing in Vietnam. With more young people educated about biodiversity, the crisis sub-theme can elicit responses from them more in line with how ENGO personnel feel about the threat to rhino survival. Change Director Hoang Thi Minh Hong said:

> [At] international school we talk about biodiversity because they are intellectual . . . We do [use] some crisis [frames] with elites and intellectual classes of grown-ups like [the] ones who are more civilized.
> (Personal communication, November 20, 2016)

The goal of the brutality sub-theme in press releases was to make consumers mentally connect rhino horn with the animal's death. WildAid's Managing Director John Baker said the "When the buying stops, the killing can too" slogan of a previous WildAid campaign in China to reduce demand for ivory and shark fin came from baseline research showing consumers there have an existing conservation attitude but were unaware that animals died to supply the products. Baker said the situation was the same with rhino horn consumers in Vietnam, so it applied the same slogan in the rhino horn campaign: "No one knew the elephant was dying to get that. You explain, you inform them of that stuff, and their attitude would instantly change to 'Oh, I would never buy that if I knew it's killing an animal'" (personal communication, November 17, 2016).

Change, WildAid's Vietnamese partner, followed WildAid's lead with the brutality sub-theme by using the same message, but Change's media relations officer, Tung Nguyen Trang, said it wasn't culturally suited because of utilitarian views of wildlife:

> The second [message] is rhino would be killed for his horn . . . We want to focus on the emotion. We want the public to see that when the rhino is poached it makes [the public] feel bad about this. When they buy rhino horn, when they buy just a piece of rhino horn, and just ask them – it will link to a death of a rhino. . .? I think it [does] not have a good response from the public about this message, yeah, because the death of rhino is just like a death of a chicken.
>
> (Personal communication, July 2, 2016)

Vice Director of Education for Nature Vietnam (ENV) Nguyen Thi Phuong Dung suggests, however, that the brutality sub-theme may be effective with young people: "It [graphic photographs] is more about the young people. It's more about the welfare issue. Because it is so graphic, people think it is horrible. As a human being, you stop doing that to the rhino" (personal communication, November 18, 2016).

Of the remaining five sub-themes, the chapter will discuss corporate social responsibility, because it features strongly in the campaign of TRAFFIC, which is arguably the "leading non-governmental organization working globally on trade in wild animals and plants" (TRAFFIC, n. d.) that advises and supports enforcement of CITES. TRAFFIC launched the "Chi" campaign in September 2014. It is based on the concept of "Chi" ("Will") within Vietnamese culture, signifying the power of what lies within (TRAFFIC, 2014, September 22), targeting wealthy businessmen-consumers with its positive message that "success, masculinity and good luck flow from an individual's internal strength of character" (TRAFFIC, 2014, September 22), not from a piece of rhino horn.

McWilliams (2015) defined corporate social responsibility as "abstract actions of firms that contribute to social welfare, beyond what is required for profit maximization." Corporate social responsibility represents corporations' duty to protect society (Holme & Watts, 1999). TRAFFIC's presentation at the 2016 Hanoi Illegal Wildlife Trade Conference discussed how the organization linked its "Chi" campaign to the emerging area of corporate social responsibility for Vietnamese companies through partnering with the Vietnamese

Chamber of Commerce and Industry (VCCI). The "Chi" campaign targets wealthy businessmen and upper-middle class citizens, whom a 2013 survey identified as "the strongest driver of the current rhino poaching crisis" (TRAFFIC, 2013, p. 1). The survey revealed that women in their 50s were the main buyers of rhino horn, while end users were wealthy aspirational men over 40 (Truong et al., 2016, p. 356). TRAFFIC partnered with VCCI to target this consumer group. Then, through a series of workshops, together they found that corporate social responsibility was an aspect of global commerce that Vietnamese businesses wanted to embrace. Hence, they decided to leverage the "Chi" campaign by suggesting that companies incorporate zero tolerance of illegal wildlife trade into their corporate social responsibility policies.

Through this aspect of the campaign, TRAFFIC also aimed to reach and work directly with business leaders and recruit some as messengers or ambassadors. TRAFFFIC produced press releases about the workshops, which highlighted that international reputation and legal risk were potential consequences for any businessman sharing or gifting rhino horn.

TRAFFIC partnered and workshopped with VCCI before deciding to incorporate corporate social responsibility into the "Chi" campaign. VCCI's Le Thi Thu Thuy's (personal communication, November 17, 2016) comments suggest that brainstorming with partners was an ENGO routine (Scheufele, 1999) that influences production of frames. It is also significant that most of the key people who took part in the brainstorming, including Le Thi Thu Thuy and TRAFFIC's Vietnam Representative Trinh Nguyen, were Vietnamese and hence knowledgeable about Vietnamese culture and business culture. Le explained how TRAFFIC increased the persuasiveness of the Responsibility frame by including the legal risks of illegal wildlife trade. Le said:

> The key motivation for business to tackle wildlife crime is to avoid risk. Adopting zero tolerance against wildlife can help a business mitigate legal and reputational risk. Employees are learning about the more stringent law and harsher punishment of wildlife crime that will come into effect in 2017 under the penal code. Under this code it is illegal to consume, keep, buy, sell, or trade protected wildlife. So, the best way to avoid becoming a wildlife criminal is to adopt zero tolerance of illegal wildlife consumption in your public and private life.
>
> (Personal communication, November 17, 2016)

Discussion and conclusion

The sub-themes that comprise the Responsibility frame can be classified based on how pertinent they are to the audience (Cox, 2010; Maibach, Nisbet, Baldwin, Akerlof, & Diao, 2010). For ENGOs, the pressing issue with reducing demand in Vietnam is the crisis and brutality facing rhinos in Africa, but there are problems in making the plight of African rhinos pertinent in Vietnam. By using the concepts of communicator and audience frames (Entman, 1993;

Scheufele, 1999), ENGOs tend to frame responsibility in a way that is pertinent to the ENGOs themselves, rather than to the Vietnamese. However, through adopting various routines, ENGOs are engaged in "organizational learning" (Princen, 1994, p. 150) in their framing and re-framing processes. Hence, they are developing frames, such as the Corporate Social Responsibility sub-theme, which are more pertinent and thus more persuasive to their target audience. This is a leveraging strategy that Cox (2010) argues goes beyond framing environmental communication because it forms a bridge from merely raising awareness to motivating changes in behavior. It is also important for ENGOs to include evidence of Vietnamese consumption and involvement in the Responsibility frame to make poaching in South Africa pertinent to the Vietnamese media and government.

To effectively reduce demand for illegal wildlife trade, behavior-change campaigns must be evidence-based (Burgess, 2016). ENGOs are tied to evidence (Anderson, 1991; Hansen, 1993) gleaned from their own research that supports their environmental goals. Research is also an important criterion for funding (World Bank, 2016). In 50 of the 52 press releases featuring the Responsibility frame, ENGOs framed Vietnam as responsible by ambiguously associating the consequences of poaching to demand and consumption there. That placed and associated the two concepts together symbolically (Entman, 1993; Blood, Putnis, & Pirkis, 2002). The ENGOs imposed causal themes, statements, and rhetorical structures, which journalists tend to use (Pan & Kosicki, 1993), but without including evidence. They fulfilled a key function of framing – "causal interpretation" (Entman, 1993) – but excluded available evidence at the expense of the legitimacy of their campaign (Anderson, 1991; Hansen, 1993).

The importance of including evidence of consumption and trade in Vietnam in the form of statistics was highlighted when Newspaper Editor A erroneously said there was little evidence that Vietnam was the key consumer nation. As all media in socialist/communist Vietnam is government-owned (Pham et al., 2017), the editor's views may be influenced by his ideological and political orientations and organizational pressures and constraints (Scheufele, 1999). Pham et al. (2017) argued that the Vietnamese media tend to reflect the government view. The editor's view is also potentially supported by Vietnam's long history of trading wildlife to China (Van Song, 2008). While African poaching statistics can be evidence of Vietnamese demand (UNODOC, 2020), the Vietnamese public and media personnel may not perceive these statistics as evidence. Providing statistical evidence of high levels of consumption and illegal trade there in press releases may help change the views of news editors, government officials, and the public. However, there is a possibility that this approach could backfire and actually increase interest in rhino horn by creating the perception that everyone uses it.

Gitlin (1980) found that framing is achieved through persistent patterns of selection, emphasis, and exclusion. It could be argued that persistent patterns in press releases often relate to routine. It is plausible that finding and citing evidence was not part of the ENGOs' routine in preparing press releases. If it were, then they likely would have cited an important research paper (Truong et al., 2016) describing the significant amount of illegal rhino horn consumption by men in positions of wealth and power in middle-class areas of Hanoi and Ho Chi Minh City. This research paper found that about half of the men surveyed (287 out of 608) admitted consuming rhino horn.

ENGOs dealt with cultural challenges and a tendency towards self-centrism through strategies that represent organizational learning (Princen, 1994). To develop cross-cultural framing, they consulted local staff, family, and friends on cultural issues. Lakoff (2010) argues for a need to build environmental background frames. This is happening with students in Vietnam, whom ENGOs engage with at schools and universities about conservation. While ENGOs identified problems with the crisis and brutality sub-themes with rhino horn consumers and the general public, they reported greater success with students. The ENGOs viewed educated young Vietnamese as an ideological bridge in society to older consumers of rhino horn. The ENGOs tended to focus less on their own views and more on the Vietnamese public and what drives consumption by developing effective routines, such as employing local staff, seeking local feedback, conducting social research, holding workshops, and partnering with local ENGOs and business groups. Such was the case with TRAFFIC's partnership with VCCI, which led to an understanding and successful application in their framing of a new important aspect of business in Vietnam.

The results outlined in this chapter suggest implications for further research, such as a framing study of Asian ENGOs' social media outputs aimed to reduce demand for rhino horn or other contraband. This research could be coupled with a framing analysis of social media and websites where rhino horn or other wildlife contraband such as ivory are advertised. In Vietnam, which ranks 15th in the world for social media use (Digital Marketing Institute, 2017) and is an international hub for illegal wildlife consumption and trade, such a study may provide a greater understanding of communication efforts both opposing and promoting one of the greatest drivers of species extinction, the illegal wildlife trade.

Acknowledgments

This chapter draws on research previously presented by Smith, M. (2019). ENGO framing of Vietnam's responsibility in the illegal rhino horn trade. International Environmental Communication Association, Conference on Communication and Environment.

Interviewees

Change Director – Hoang Thi Minh Hong.
Change Media Relations Officer – Tung Nguyen Trang.
ENV Vice Director – Nguyen Thi Phuong Dung.
Newspaper editor A (Chose to remain anonymous)

Traffic Vietnam representative –Trinh Nguyen.
Vietnam Chamber of Commerce (VCCI) Small and Medium-sized Enterprises Development Centre Vice Director – Le Thi Thu Thuy.
WCS Policy Coordinator – Duong Viet Hong.
WildAid Managing Director – John Baker.

References

Anderson, A. (1991). Source strategies and the communication of environmental affairs. *Media, Culture and Society, 13*(4), 459–476.

Anderson, A. (1997). *Media, culture and the environment*. London: University College London Press.

Beachey, R. W. (1967). The East African ivory trade in the nineteenth century. *Journal of African History, 8*(2), 269–290.

Blood, R. W., Putnis, P., & Pirkis, J. (2002). Mental-illness news as violence: A news frame analysis of the reporting and portrayal of mental health and illness in Australian media. *Australian Journal of Communication, 29*(2), 59–81.

Burgess, G. (2016). Powers of persuasion? *TRAFFIC Bulletin, 28*(2), 65–68. Retrieved from www.traffic.org/site/assets/files/3385/powers-of-persuasion.pdf

Cheung, H., Mazerolle, L., Possingham, H. P., & Biggs, D. (2018). Medicinal use and legalized trade of rhinoceros horn from the perspective of traditional Chinese medicine practitioners in Hong Kong. *Tropical Conservation Science, 11*(1–8). doi:10.1177/1940082918787428

Cox, R. (2010). Beyond frames: Recovering the strategic in climate communication. *Environmental Communication, 4*(1), 122–133.

Cox, R., & Pezzullo, C. (2016). *Environmental communication and the public sphere*. Thousand Oaks, CA: Sage.

de Monestrol, H. (1952). *Hunting and wildlife of Indochina*. Saigon: A. Portail.

Digital Marketing Institute. (2017). Social media: What countries use it most & what are they using? Retrieved from https://digitalmarketinginstitute.com/blog/social-media-what-countries-use-it-most-and-what-are-they-using

Dirikx, A., & Gelders, D. (2009). Global warming through the same lens: An exploratory framing study in Dutch and French newspapers. In T. Boyce & J. Lewis (Eds.), *Media and climate change* (pp. 200–210). Oxford, UK: Peter Lang.

Drury, R. (2009). Hungry for success: Urban consumer demand for wild animal products in Vietnam. *Conservation and Society, 9*(3), 247–257.

Entman, R. M. (1993). Framing: Toward clarification of a fractured paradigm. *Journal of Communication, 43*(4), 51–58.

ENV. (2014, September 8). Vietnamese delegation arrives in South Africa on mission to help protect rhinos (Press Release). Retrieved from https://env4wildlife.org/2014/09/08/vietnamese-delegation-arrives-in-south-africa-on-mission-to-help-protect-rhinos/

Environmental Investigation Agency. (2016). Time for action: End the criminality and corruption fueling wildlife crime. Retrieved from https://eia-international.org/wp-content/uploads/EIA-Time-for-Action-FINAL-1.pdf

Fowlds, W. (2015, June 24). *Vet describes emotional toll of responding to brutal rhino poaching* (Video). Retrieved from https://news.mongabay.com/2015/06/video-vet-describes-emotional-toll-of-responding-to-brutal-rhino-poaching/

Gamson, W. A., & Modigliani, A. (1989). Media discourse and public opinion on nuclear power: A constructionist approach. *American Journal of Sociology, 95*(1), 1–37.

Gitlin, T. (1980). *The whole world is watching: Mass media in the making and unmaking of the New Left*. Berkeley, CA: University of California Press.

Hansen, A. (1993). Greenpeace and the press coverage of environmental issues. In A. Hansen (Ed.), *The mass media and environmental issues* (pp. 150–178). Leicester, UK: Leicester University Press.

Hansen, A. (2011). Communication, media and environment: Towards reconnecting research on the production, content and social implications of environmental communication. *International Communication Gazette*, *73*(1–2), 7–25.

Harper, J. (2015). $60K a pound: Illegal rhino horn now declared more valuable than gold, diamonds and cocaine. Retrieved from www.washingtontimes.com/news/2015/may/17/rhino-horn-considered-cure-all-and-aphrodesiac-now/

Holme, R., & Watts, P. (1999). *Corporate social responsibility*. Geneva, Switzerland: World Business Council for Sustainable Development.

Jeffreys, E. (2016). Translocal celebrity activism: Shark-protection campaigns in mainland China. *Environmental Communication*, *10*(6), 763–776.

Knight, M. (2017). African rhino specialist group report. Retrieved from www.rhinoresourcecenter.com/pdf_files/151/1517207949.pdf

Lakoff, G. (2010). Why it matters how we frame the environment. *Environmental Communication*, *4*(1), 70–81.

Lester, L. (2007). *Giving ground: Media and environmental conflict in Tasmania*. Hobart, Australia: Quintus Publishing.

Lin, T. T. (2012). Cross-platform framing and cross-cultural adaptation: Examining elephant conservation in Thailand. *Environmental Communication*, *6*(2), 193–211.

Maibach, E. W., Nisbet, M., Baldwin, P., Akerlof, K., & Diao, G. (2010). Reframing climate change as a public health issue: An exploratory study of public reactions. *BMC Public Health*, *10*(1), 299–310.

McWilliams, A. (2015). Corporate social responsibility. In C. L. Cooper, J. McGee, & T. Sammut-Bonnici (Eds.), *Wiley encyclopedia of management*. Chichester, UK: John Wiley & Sons.

Milliken, T. (2014). Illegal trade in ivory and rhino horn: An assessment report to improve law enforcement under the wildlife TRAPS project. Retrieved from www.usaid.gov/sites/default/files/documents/1865/W-TRAPS-Elephant-Rhino-report.pdf

Milliken, T., & Shaw, J. (2012). The South Africa – Viet Nam rhino horn trade nexus: A deadly combination of institutional lapses, corrupt wildlife industry professionals and Asian crime syndicates. Johannesburg, South Africa. Retrieved from www.trafficj.org/publication/12_The_SouthAfrica-VietNam_RhinoHorn_Trade_Nexus.pdf

Mol, A. P. (2009). Environmental governance through information: China and Vietnam. *Singapore Journal of Tropical Geography*, *30*(1), 114–129.

Mukpo, A. (2020). As calls to shutter wildlife markets grow, China struggles with an industry worth billions. Retrieved from https://news.mongabay.com/2020/04/as-calls-to-shutter-wildlife-markets-grow-china-struggles-with-an-industry-worth-billions/

Pan, Z., & Kosicki, G. M. (1993). Framing analysis: An approach to news discourse. *Political Communication*, *10*(1), 55–75.

Pham, T. T., Gregorio, M., & Brockhaus, M. (2017). REDD+ politics in the media: A case study from Vietnam. *International Forestry Review*, *19*(1), 69–80.

Princen, T. (1994). NGOs: Creating a niche in environmental diplomacy. In T. Princen & M. Finger (Eds.), *Environmental NGOs in world politics: Linking the local and the global* (pp. 29–47). London: Routledge.

Reporters sans Frontières. (2021). World press freedom index. Retrieved from https://rsf.org/en/ranking

Save the Rhino. (n.d.). *Poaching statistics*. Retrieved from www.savetherhino.org/rhino-info/poaching-stats/

Scheufele, D. A. (1999). Framing as a theory of media effects. *Journal of Communication, 49*(1), 103–122. doi:10.1111/j.1460-2466.1999.tb02784.x

Semetko, H. A., & Valkenburg, P. M. (2000). Framing European politics: A content analysis of press and television news. *Journal of Communication, 50*(2), 93–109.

Smith, M. S. (2015a, March 2). The Saigon horn. Retrieved from https://micsmithgeographic.wordpress.com/2015/03/02/the-saigon-horn/

Smith, M. S. (2015b, June 5). Amid rhinoceros poaching frenzy, dark days for South African society. Retrieved from https://news.mongabay.com/2015/06/amid-rhinoceros-poaching-frenzy-dark-days-for-south-african-society/

Smith, M. S. (2017). The framing of rhino horn demand reduction by Vietnamese ENGOs: Cultural and other factors influencing the "responsibility" frame in ENGO media outputs (Master's thesis). Griffith Research Repository. Retrieved from http://hdl.handle.net/10072/371949

Smith, M. S. (2018a). Framing rhino horn demand reduction in Vietnam: Dismissing medical use as voodoo. *Pacific Journalism Review, 24*(2), 241–256.

Smith, M. S. (2018b). Credibility and trust issues stemming from an ambiguous science frame: Reducing demand for rhino horn in Vietnam with the fingernail metaphor. Paper presented to Iowa State University Summer Symposium on Science Communication. Understanding the Role of Trust and Credibility in Science Communication, Ames, Iowa. doi:10.31274/sciencecommunication-181114-7

South Africa Department of the Environment, Forestry, and Fisheries [DEFF]. (2019). Department of Environment, Forestry, and Fisheries reports back on rhino poaching in first six months of 2019. Retrieved from www.environment.gov.za/mediarelease/barbaracreecy_rhinopoaching_sixmonthreportback_july2019

South Africa Department of the Environment, Forestry, and Fisheries [DEFF]. (2021). Department of Environment, Forestry, and Fisheries report back on rhino poaching in South Africa in 2020. Retrieved from https://www.gov.za/speeches/environment-forestry-and-fisheries-rhino-poaching-south-africa-2020-1-feb-2021-0000#

Tankard, J. W. (2001). The empirical approach to the study of media framing. In S. Reese, O. Gandy, & A. Grant (Eds.), *Framing public life: Perspectives on media and our understanding of the social world* (pp. 95–106). London: Routledge.

TRAFFIC. (2013). Rhino horn consumers who are they? Retrieved from www.traffic.org/site/assets/files/8094/rhino-horn-consumers-who-are-they.pdf

TRAFFIC. (2014, September 22). Innovative campaign promotes success from within (Press Release). Retrieved from www.traffic.org/news/innovative-campaign-promotes-success-from-within/

TRAFFIC. (2015, June 15). Vietnam Chamber of Commerce and Industry encourages social and environmentally responsible business practices (Press Release). Retrieved from www.traffic.org/news/vietnam-chamber-of-commerce-and-industry-encourages-social-and-environmentally-responsible-business-practices/

TRAFFIC. (n.d.). Our mission. Retrieved from www.traffic.org/about-us/our-mission/

Truong, V. D., Dang, N. V. H., & Hall, C. M. (2016). The marketplace management of illegal elixirs: Illicit consumption of rhino horn. *Consumption Markets & Culture, 19*(4), 353–369. doi:10.1080/10253866.2015.1108915

UNODOC. (2016). World wildlife crime report: Trafficking in protected species. Retrieved from www.unodc.org/documents/data-and-analysis/wildlife/World_Wildlife_Crime_Report_2016_final.pdf

UNODOC. (2020). World wildlife crime report: Trafficking in protected species. Retrieved from www.unodc.org/documents/data-and-analysis/wildlife/2020/World_Wildlife_Report_2020_9July.pdf

Van Song, N. (2008). Wildlife trading in Vietnam: Situation, causes, and solutions. *The Journal of Environment & Development, 17*(2), 145–165.

WCS. (2014, November 3). Has demand for rhino horn truly dropped in Vietnam (Press Release). Retrieved from https://blog.nationalgeographic.org/2014/11/03/has-demand-for-rhino-horn-truly-dropped-in-vietnam/

Wells-Dang, A. (2010). Political space in Vietnam: A view from the "rice-roots." *The Pacific Review, 23*(1), 93–112.

WildAct. (2014, October 16). World youth rhino summit post activity: Vietnamese delegate Tuan Nam calling his friends to speak out for rhinos (Press Release). Retrieved from www.wildact-vn.org/news?offset=1412170044317&reversePaginate=true

WildAct. (2016, June 20). Dispelling the myth: Protecting human and rhinos? Pt 1(Press Release). Retrieved from www.wildact-vn.org/news?offset=1494707436060

WildAct.(2016, November 17). Critically endangered species for sale online on Facebook in Vietnam (Press Release). Retrieved from www.wildact-vn.org/news?offset=1424058387476&reversePaginate=true

WildAid.(2016, September 22). In Vietnam, building religious support to save rhinos (Press Release). Retrieved from https://wildaid.org/in-vietnam-building-religious-support-to-save-rhinos/

World Bank. (2016). New analysis shows scale of international commitment to tackle illegal wildlife trade: Over 1.3 billion since 2010. Retrieved from www.worldbank.org/en/news/press-release/2016/11/17/new-analysis-shows-scale-of-international-commitment-to-tackle-illegal-wildlife-trade-over-13-billion-since-2010

WWF. (2016, November 18). WWF reaction to the Hanoi statement on wildlife crime (Press Release). Retrieved from https://wwf.panda.org/?285012/WWF-Reaction-to-the-Hanoi-Statement-on-Wildlife-Crime

Yang, G., & Calhoun, C. (2013). Media, civil society and the rise of the green public sphere in China. In L. Lester & B. Hutchins (Eds.), *Environmental conflict and the media* (pp. 272–289). New York: Peter Lang.

Part III

Media emphasize the charismatic

Ignoring the rest

9 Front pages are for the charismatic

The case of the cute giant panda

Ryan Wallace

In 1936, American fashion designer and New York socialite Ruth Harkness ventured to Tibet in search of her former husband's legacy (Croke, 2006; Li, 2020). After a successful adventure to Indonesia where he captured a live Komodo dragon, Ruth's husband William Harkness died on an expedition to China attempting to bring back the first-ever live giant panda (*Ailuropoda melanoleuca*) (Croke, 2006). However, it was Ruth who emerged from the dense bamboo forests and harsh terrain of China's border with Tibet to introduce the West to its first panda cub: Su Lin (Croke, 2006). Translated as "a little bit of something precious," Su Lin became an international phenomenon as newspapers around the world told the story of the cub that made it to the West (Bowes & Gee, 2013; Croke, 2006). No animal had ever received the same level of international attention in the history of modern media, and Su Lin became the first of many pandas to capture the hearts of mass audiences (Buckingham, David, & Jepson, 2013; Croke, 2006; Li, 2020).

With Su Lin's arrival in the United States – listed as a dog in the customs papers that allowed it to leave China – the species garnered global attention for its charm. Their rarity and unique appearance, once shielded by almost entirely impenetrable bamboo forests, made them popular among the masses, who flocked to zoos and newsstands just to catch a glimpse or read the latest about them (Croke, 2006; Li, 2020). A number of entities have strategically used the mass appeal of the giant panda for their own purposes since the latter half of the 20th century. Most notably, the Chinese government has actively positioned the species as a national emblem. Mao Zedong, the communist revolutionary who founded the People's Republic of China after World War II, first popularized the practice of gifting and loaning giant pandas to other countries (Bowes & Gee, 2013; Buckingham et al., 2013). This international "panda diplomacy" contributed to building strategic friendships and alliances with foreign nations, while at the same time emphasizing the rarity of the species and the challenges it faced in the wild (Buckingham et al., 2013). Portrayed by China as a peaceful and soft icon during its global rise to power, the exchange of captive pandas was used for clear political purposes (Huang & Wang, 2020; Nicholls, 2012).

The World Wide Fund for Nature (WWF) used the giant panda as its logo, making the panda an icon for conservation efforts and raising global attention and concern for their protection in their natural habitats (Li, 2020; Lousley, 2016). Although some biologists and conservation organizations have openly criticized the vast funding allocated to giant panda conservation in recent decades, diverting attention and vital resources from the conservation of biodiversity, the species has a proven track record as both an ambassador and fundraiser (Buchen, 2008). In China, where environmental practices have long been criticized, protecting the forest ecosystems of the giant pandas has helped save hundreds of other vertebrate species and thousands of insect species (Buchen, 2008).

As important characters in conservation discourse, species like the giant panda are common tropes for the mass media to use in their construction of news regarding a wide variety of ecological topics, ranging from endangered species to anthropogenic impacts on the environment. However, while communicating such issues is vital, these topics have largely become entangled with political, sociocultural, and scientific dimensions (Born, 2019). The use of iconic species such as giant pandas and polar bears facilitates mass communication of these issues and their significance in contemporary discourse (Born, 2019). These icons not only embody adverse changes to nature, but also provide mass media with a way of framing these issues for the general public with a familiar "face" in these stories (Born, 2019; Slocum, 2004).

With the power to influence not only mass audiences, but also public and policy agendas that impact conservation planning and management, mass media serve as an important subject of analysis for their agenda-setting role (Maxwell, 2004; McCombs & Shaw, 1972). To successfully implement conservation intervention strategies, plans must be far more than ecologically effective – they must also take into account economic, sociocultural, and even political dimensions of diverse groups of stakeholders (Treves, Wallace, & White, 2009). By helping shape the political and social conditions needed for conservation policies to succeed, media coverage of endangered species can draw attention to vulnerable species, affect the public's attitudes, and lead to actionable change (Gusset et al., 2008; Treves et al., 2009). At the same time, much of this media coverage is not driven by the importance of the information or the relevance of particular species. Rather, it is driven by media professionals whose coverage often reflects existing schema and frames (Jacobson, Langin, Carlton, & Kaid, 2012; Jacobson et al., 2019). This chapter analyzes one such aspect of the existing media logic, the concept of charismatic megafauna, to better understand the value and challenges of using iconic species for global conservation discourse (Table 9.1). Instead of serving particularly important ecological roles like their counterpart keystone species, species like charismatic megafauna are selected for their ability to gather support from diverse audiences and galvanize action (Jepson & Barua, 2015). While journalists and conservation organizations may use species like the giant panda to garner public support and build on existing conservation discourse, this coverage can contribute to hegemonic

Table 9.1 Key focus of the chapter with some of the most widely circulated publications in the United States and the six most charismatic species of megafauna.

Publications	*Species of Charismatic Megafauna*
Boston Globe	Elephant
Chicago Tribune	Giant panda
Discover Magazine	Giraffe
Los Angeles Times	Lion
National Geographic	Panther
Nature Magazine	Tiger
New York Daily News	
New York Times	
New York Post	
Popular Science Magazine	
San Francisco Chronicle	
Science Magazine	
Scientific American	
Smithsonian Magazine	
Economist	
Time Magazine	
USA Today	
Wall Street Journal	
Washington Post	

representations of species that are endangered or threatened. It can also complicate the economic, sociocultural, and even political dimensions of conservation discourse.

The giant panda is among the top 10 most charismatic and well-known animal species, along with several prominent "big cat" species, elephants, and giraffes (Albert, Luque, & Courchamp, 2018). With a large, round face, a clumsy nature, and a fluffy coat, the giant panda possesses childlike characteristics (Huang & Wang, 2020). Since the first cub arrived in the West clutched in an American socialite's arms, the public's willingness to fawn over and care for the species was clear (Croke, 2006). Media attention and scientific interest subsequently drove conservation efforts to protect the species, its habitat, and biodiversity of other species alongside them (Lousley, 2016). Even though poaching and other anthropogenic activities brought the giant panda to the brink of extinction, today it is slowly rebounding (Swaisgood, Wang, & Wei, 2016). In 2016, the International Union for Conservation of Nature (IUCN) Red List downgraded it from "endangered" to "vulnerable" status, although its fate remains under significant threat. With fewer than 1,000 mature giant pandas left in the wild, a wide range of forces jeopardize their forest ecosystems throughout China. The rapid expansion of human settlements and commercial development, processes of agriculture, building, extractive industries, and many other ecological threats pose the greatest challenges in the form of habitat loss (Swaisgood et al., 2016; Swaisgood, Wang, & Wei, 2018).

The giant panda is considered the quintessential example of a charismatic megafauna for its role in mass media and scientific literature, by drawing attention to issues of conservation and endangered species (Walpole & Leader-Williams, 2002). Studies based on survey data and analyses of salience in mass media have sought to better understand what makes a species "charismatic" to fill the conceptual definitions missing in conservation research (Albert et al., 2018). However, in evaluating the use of particular species to raise funds as surrogates for conservation efforts and to obtain the backing of policy makers and the public, it is clear that charisma is simply an attribute of an already established type of surrogate called "flagship species" (Albert et al., 2018).

Built on the notion that select species can capture public attention and foster support for conservation efforts, flagship species are conceptually defined as high-profile, charismatic species used to raise public awareness and funding for conservation (Walpole & Leader-Williams, 2002). In practice, however, the species has proven to serve a greater socioeconomic than ecological role in these efforts (Walpole & Leader-Williams, 2002). Adopted by the WWF as its icon in the 1960s, the giant panda is one of the world's most widely recognized species despite its rarity in the wild (Nicholls, 2011, 2012). As a flagship species, it has amassed a global following and garnered support for conservation efforts. While the scale of funds that the species has raised remains uncalculated, analyses of the "pandanomics" behind their conservation and reserves reveal that it generates billions more dollars than it costs to protect them (Forbes, 2018; Wei et al., 2018). With the most current estimates from 2010 suggesting that their nature reserves generate between $2.6 and $6.9 billion in revenue each year, the giant panda yields between 10 and 27 times the costs of protecting its habitat and the species as a whole (Wei et al., 2018). More important than mere pandanomics are the practical implications of their time in the spotlight.

As they have ascended to global prominence and captured the hearts and the pocketbooks of donors, they have also facilitated changes in conservation at a broader level. The iconization of giant pandas by organizations like the WWF has expanded the focus on charismatic megafauna to encompass threatened ecosystems and endangered species more generally (Lousley, 2016). After opening relations to the West and building connections with conservation organizations, the Chinese government implemented important strategies to facilitate the species' survival (Buckingham et al., 2013; DeMarco, 2015; Li, 2020; Wei et al., 2018). China has implemented two important policies, the Natural Forest Conservation Program and the Grain-to-Green Program, which not only help conserve the giant panda's natural habitat ranges, but also make it possible for the government to establish national nature reserves dedicated to its protection (Li & Pimm, 2016; Ren et al., 2015). Conservation planning, as a complex process that often accounts for different goals and many species apart from those well known for their charisma, has relied on proxies like the giant panda to describe, characterize, and measure aspects of progress and overall biodiversity (Di Minin & Moilanen, 2014). While charismatic megafauna are not perfect stand-ins for threatened biodiversity, their effective marketing value can facilitate vital policies and funding that make conservation practices possible, and by sharing the spotlight with other species, charismatic megafauna can bring other threats to the forefront of conservation discourse.

Research questions

Focusing on the giant panda, this study sought to better understand how certain species have become engrained with conservation and environmental discourse, becoming icons for nations, organizations, and environmental movements. Other charismatic megafauna also have become popular subjects in mass media news stories, often associated with climate change and ecological frames (Boykoff, 2008). However, their prominence in mass media varies as journalists determine the importance and relevance of particular species in their storytelling (Jacobson et al., 2012). Understanding the prevalence and context of this discourse is important because news coverage not only can impact the public understanding of ecosystems and the ongoing processes of extinction, but also their perceptions about conservation policies (Jacobson et al., 2012). This agenda-setting function is a cornerstone of mass media and highlights its important role in the development of successful conservation strategies. To implement an effective strategy, conservation planners not only require favorable ecological conditions, but also sociocultural and political support (Treves et al., 2009).

By looking at different forms of news media – text articles and cover pages from some of the United States' most widely consumed publications – this study examined multimodal forms of discourse to better understand not only trends in the texts but also meaning-making on the covers of influential publications. While articles provided a fundamental understanding of conservation discourse more broadly, magazine and newspaper cover pages served as the most important objects of analysis because they reflect insights into the publications and serve as complex sources of visual–verbal rhetoric that persuade and entertain audiences (Clucas, McHugh, & Caro, 2008; Held, 2005). At its core, this chapter addresses two fundamental research questions:

RQ1: Compared to other charismatic megafauna, to what extent is the giant panda a central part of discourse on endangered species and conservation?

RQ2: What is the importance of having giant pandas on the covers of mass media publications?

Methods

To answer these research questions, this study used a mixed-methodological approach to both analyze the content of conservation and endangered species discourse and to interpret its meaning. The methods were a content analysis for articles related to the topic and a multimodal discourse analysis for covers showing charismatic megafauna. Articles and covers were gathered through three databases (Factiva, Kiosko.net, and Gale Virtual Library), as well as from each publication's archives for 19 of the United States' most highly circulated publications. They included top daily newspapers, popular science magazines, and other periodicals.[1]

A corpus of articles was gathered ($n = 335{,}451$) using key search terms "conservation" and "endangered species" from January 1, 1980, through July 31, 2020. Articles were coded to determine whether they mentioned certain species of charismatic megafauna in reference to either key term. This content analysis provided the foundation for the comparison of giant pandas with other charismatic species.

Front pages and magazine covers of issues mentioning and visualizing charismatic megafauna were also gathered from these publications, focusing on the six most charismatic species: elephants, giraffes, lions, giant pandas, panthers, and tigers (Albert et al., 2018). In total, a corpus of $n = 86$ covers was gathered relevant to these species. These covers served as the foundation for the multimodal discourse analysis, which sought to explain how and why giant pandas and other species have become popular icons within the conservation frame commonly used in existing media logic.

Findings

To answer RQ1 (*Compared to other charismatic megafauna, to what extent is the giant panda a central part of discourse on endangered species and conservation?*) all "conservation" and "endangered species" articles were coded for these key terms, in addition to the names of the six most charismatic megafauna species. Among "endangered species" articles ($n = 50{,}083$), statistically significant differences ($p = 0.041$) were indicated in the discourse with regards to the frequency of which certain charismatic megafauna were mentioned versus others. ANOVA analyses found differences in the means, with elephants being the most widely discussed ($\mu = 129.8$) in endangered species discourse across all 19 publications and giraffes being the least ($\mu = 21.33$). Giant pandas were the fourth most widely discussed ($\mu = 49.3$), but no statistically significant difference was found among the other top three species ($p = 0.437$).

Among the broader "conservation" articles ($n = 285{,}368$), a similar pattern existed in the data. ANOVA analyses indicated a statistically significant difference ($p = 0.034$) in the mean number of articles each species was mentioned in, this time with lions being the most widely discussed ($\mu = 351$) and again giraffes being the least ($\mu = 47.7$). In discourse about conservation, giant pandas ($\mu = 79.9$) once again came in fourth among the most charismatic species. Additional ANOVA analyses indicated that among giant pandas and the other top three species, no statistically significant difference existed ($p = 0.242$) between the means.

In contrast, analyses of the covers ($n = 86$) of the 19 publications found a different hierarchy of species mentioned or visualized on the front pages of popular US newspapers and magazines. ANOVA analyses indicated a statistically significant difference ($p = 0.002$) in the number of publications that each species was mentioned in. The most popular species, seen on the cover of 13 publications, was the giant panda. The least popular was the panther, which appeared on the cover of only one publication.

While there is a division among species of charismatic megafauna that are and are not widely discussed in discourse of "endangered species" and "conservation," the results of RQ1 suggest that giant pandas are among the most widely discussed, despite being mentioned in fewer articles. They still represent an important component of conservation discourse and occupy a particularly pervasive role on the covers of news publications. Their presence extends far beyond popular science publications, like *Scientific American* and *National Geographic*, to popular local newspapers, several newspapers of record including the *New York Times* and *Washington Post*, and even to popular business and economics publications like the *Wall Street Journal* and *Economist*.

To answer RQ2 (*What is the importance of having giant pandas on the covers of mass media publications?*) all covers that focused on charismatic species were analyzed and interpreted to account for textual and visual forms of meaning-making used in magazine and newspaper cover designs. Given the nature of these objects of analysis, borrowing from various symbol systems to create meaning, the qualitative method of multimodal discourse analysis was determined to be the best approach to understanding the characteristics and significance of using these species as icons for conservation discourse (Kress, 2013; Kress & Van Leeuwen, 1996). The covers were coded through repeated rounds of analysis, which facilitate a thick description of the data and a systematic theoretical interpretation (Glaser, 1965; Glaser & Strauss, 2017).

As in the quantitative analyses of cover pages, species like panthers and giraffes were among the least widely discussed and visualized, both on covers and within the texts. However, even among the least discussed of the six charismatic species this chapter focuses on, the complexities of geopolitical discourse complicate the primarily ecological narrative. This is best illustrated by the wide variety of *Economist* covers. They have used giraffes to represent "Aspiring Africa," tigers to represent India's e-commerce, a toothless lion to represent the weakened West, and flying pandas to represent China's economic reforms. The *Economist* most clearly illustrates the complex ways in which discourse about iconic species are complicated by other sociocultural meanings through images, illustrations, and allegories. And it's not the only publication to use these species to represent global geopolitics. *Time* covers have also used elephants to represent India as the world's largest democracy and lions to represent the West's involvement in the Middle East and the Indian subcontinent. One 1946 *Time* cover story, "Mohamed Ali Jinnah: His Muslim tiger wants to eat the Hindu cow," is noteworthy in that it not only used two of these endangered species in its visual discourse (lions and tigers), but also because of how it played on different symbol systems. Coming at a time of intense conflict between Hindu and Muslim communities leading up to the foundation of Pakistan, this cover illustrated how endangered species like Bengal tigers served as icons for discourse even early on, far beyond conservation – including religion and complex geopolitics (Hamdani, 2016).

As for the endangered species narrative, this frame served an important role in focusing stories and covers on conservation discourse. This narrative included cover headlines like the *Los Angeles Times*' "A Black-and-White Effort" discussing China's program for giant panda reintroduction to the wild, *Time*'s "Doomed: Why the Regal Tiger is on the Brink of Extinction," and *Smithsonian* magazine's "The Last Giraffe: Will the Beloved Giant Disappear Before Scientists Even Begin to Understand It?" But in the absence of the endangered species narrative, the complexities of the discourse surrounding these species often ignored the precarious status of their survival. Nearly all charismatic megafauna species in this study had at least one cover that mentioned the dire state of the species or the "last" remaining individuals of their kind. However, in their visual discourse, they varied greatly between predators like lions and tigers and herbivores like giraffes, elephants, and pandas. These differences in the visual discourse can most effectively be described as a juxtaposition of predators in nature and prey in captivity. Extending as far back as the first images of Ruth Harkness and Su Lin, the visual discourse of giant pandas as illustrated by these covers is one of human intervention: giant pandas in captivity, and their dependence on humans to prevent their extinction.

Although some early covers used images of stuffed giant pandas, courtesy of US President Theodore Roosevelt's sons, most cover images of giant pandas show their lives in captivity (Croke, 2006; Grossman, 2017). From metal bars to fake rocks and metal bowls, the visual discourse focuses on cubs, mother and cub, and artificial enclosures. Even images that appear to have pandas in natural environments, such as those on the cover of *Smithsonian* and the *Washington Post*, have visual markers such as non-native species of foliage that indicate artificial environments like those found in zoos. These images and cover texts not only tap into the human–nature relationship, emphasizing panda reliance on humans for survival, but also illustrate how giant pandas have been popularized in contemporary societies and have become a species that fascinates and entertains mass audiences. From the *Washington Post*'s 2005 cover calling the National Zoo's 4-month-old cub Tai Shan "the hottest ticket in town" to the *Wall Street Journal*'s 2009 cover describing "The Whirlwind Vacation" where ecotourists traveled to China for an opportunity to kneel down and take photos with panda cubs, this discourse illustrates both the species' mass appeal and ways in which humans have obscured their own culpability in its endangerment.

Interestingly, while some cover stories openly discuss reasons for their dwindling populations – such as *Time*'s "Vanishing Act" cover from April 2009 that discussed climate change's role in "a new age of extinction," the *New York Times*' September 2019 cover that included the story "The Plight of the Captive Tiger" about Thailand's black market trade, and *Time*'s 1989 cover "The Ivory Trail" with "a story of greed and slaughter" of elephants across Asia and Africa – most covers and cover stories avoided textual and visual discourse related to many negative anthropogenic impacts on these species. For the giant panda in particular, *Wall Street Journal*, *Chicago Tribune*, and *Washington Post* covers lauded early hunting of pandas as trophies, the trade and captivity of pandas for the National Zoo, and even ecotourism where Western jet-setters encounter giant panda cubs.

It is clear that the significance of putting giant pandas on covers is largely because of their cute appearance and mass appeal, which may encourage audiences to pick these covers off of newsstands. It is also clear that economic and geopolitical issues complicate the discourse about this species and many other charismatic megafauna. In fact, conservation discourse on the covers is minimal, and at times these cute and charismatic pandas diminish attention for far more important news. This is best illustrated with a November 2019 *Washington Post* cover about National Zoo panda Bei Bei's departure "on an emotional day" taking precedence over an important section that described a turning point as "Hong Kong police moved to break the protests, and Beijing denounced a court ruling friendly to demonstrators." While the turning point story was technically higher up in the newspaper (A14), Bei Bei's departure story (B1) overshadowed other important developments throughout China. Even unintentionally, the popularity of these species can distract from other important developments. Given the nature of these covers focusing primarily on pandas in captivity, they can also pull focus from the serious threats these species face in the wild.

Discussion and conclusions

The benefit of charismatic megafauna is that they offer the public a character facing extinction to connect with emotionally (Buchen, 2008). By engaging mass audiences, they can garner financial and broader public support for conservation efforts (Buchen, 2008; McGowan et al., 2020). But if they become too central to management strategies, they may draw attention and vital resources away from lesser-known endangered species (Buchen, 2008). While the use of flagship species has proven an effective part of fundraising strategy, more research is needed to better understand if and how to prioritize flagship species in conservation strategies because prioritizing one species may impede efforts to conserve biodiversity more broadly (McGowan et al., 2020). As conservation groups have adopted these icons, researchers have begun to question the effectiveness of using them as central characters in this discourse, as well as how they may affect perceptions of nature and understandings of ecological changes (Dunaway, 2009; Manzo, 2010).

While iconic species may enable mass audiences to identify with global threats to the environment and its species, as well as provide mass media with ways of describing, localizing, and visualizing their effects, there are significant implications in using these few species as proxies for conservation discourse (Born, 2019). Use of icons can obscure ways in which humans and nature are entangled and how many of these forces relate to global concerns (Born, 2019). The covers illustrate that while some species like the giant panda may currently occupy an important role in conservation discourse, they also contribute significantly to hegemonic representations of endangered/threatened species and can unnecessarily complicate this discourse with other dimensions, such as geopolitics. Rather than advocating that specific locations or flagship species be used for conservation efforts, researchers now suggest basing the use of flagship species on broader conservation goals instead bringing more species under the umbrellas of these icons (McGowan et al., 2020). Conservation organizations, therefore, could have a larger global impact with their funding by addressing similar conservation goals in terrestrial and aquatic ecosystems (McGowan et al., 2020).

Not enough knowledge "under the umbrella"

A key criticism of the habitualized use of flagship species in conservation discourse is that focusing on "big, cute, and furry" species has led to a mammal-centric hegemony that diminishes broader conservation efforts to maintain biodiversity (Walpole & Leader-Williams, 2002). As symbols for the environment, these species become increasingly central to both local and international organizations, conservation efforts, and even news (Walpole & Leader-Williams, 2002). This analysis of US news coverage illustrated significant media attention on giant pandas, primarily because of their audience appeal. However, practical policy implications accompany their mass popularity. Because of this mass appeal, China has more than doubled the number of nature reserves dedicated to conserving pandas since 2007 and has allocated significant terrain to developing new forests for them (DeMarco, 2015; Holland, 2019). Research analyzing the geographic range of other species native to China suggests that overlaps between protected panda habitats and the natural ranges of other species can help conserve dozens of mammal, bird, and amphibian species (Li & Pimm, 2016).

However, while detailed geographical data is available for vertebrate species, there is little such data for threatened plants and invertebrates (Li & Pimm, 2016). Many other species go unnoticed while a variety of vertebrates are protected "under the umbrella of the giant panda" (Li & Pimm, 2016, p. 329). Take, for example, the fact that nearly 85% of plant species occupy only a fraction of the Earth's surface. Conservation and land protection researchers question whether it is scale and quantity or quality that matters most when preserving biodiversity (Pimm, Jenkins, & Li, 2018). Large swaths have been protected for reasons ranging from ecological to geochemical, but smaller protected areas in the right locations might serve as better refuges for many species (Pimm et al., 2018). The focus, then, should not be on protecting the largest areas of land, but prioritizing the right areas for achieving conservation goals (Pimm et al., 2018). Similarly, news coverage of ongoing conservation efforts may benefit from reporting on ecological communities rather than particular species. That way they could engage audiences with a wider diversity of species facing similar threats while avoiding geopolitical and sociocultural dimensions that detract from conservation discourse. By focusing on areas of greater biodiversity known as "hotspots," conservation efforts may develop more efficient strategies (Flather, Knowles, & Kendall, 1998). And by analyzing the geographies of threatened and endangered species and finding where their ranges and niches intersect, conservation biologists may be better able to outline management strategies and geographic regions for further protections (Flather et al., 1998).

Minding the gap

While the IUCN Red List serves as an important global standard for classifying threatened or endangered species, it often differs from national and international conservation efforts in policy and management decision-making (Harris et al., 2012). Policy strategies such as the US Endangered Species Act represent differing constraints, objectives, and priorities (Harris et al., 2012). Planning often requires not only considerations of ecological, but also economic and sociocultural dimensions. In such cases, there is a visible gap between species listed as needing protection and those whose struggles are not recognized (Harris et al., 2012). At the center of these conversations, giant pandas have distracted from these complex dimensions, but their mere existence provides reasons why mass communications should mind this gap. As complex symbols with global prominence and fascinating modern histories, they provide mass media with the ability to depict global sociocultural, geopolitical, and environmental changes. While journalists, governments, and conservationists often deemphasize the complex human–nature relationship giant pandas illustrate, they represent a much broader category of threatened species. By focusing on commonalities with other species, rather than differences, these groups may better articulate protections needed on a wider scale. This can bring more species, both charismatic and not, into conservation discourse.

Giant pandas have proven to be prime ambassadors and fundraisers. They have had significant economic and sociocultural effects. Giant pandas have also facilitated change regarding some conservation efforts, prompting nations to notice the exorbitant efforts necessary to bring a species back from the edge of extinction. However, such an emphasis on the charismatic has raised questions regarding the implications of focusing on such flagship species. The giant panda has become a central character in the semiotics of conservation discourse. The historical fascination with it may once have served a purpose in bringing together mass audiences to prevent their extinction, but as their populations rebound, the conversation should frame ongoing threats in a way that spotlights ecosystems as a whole. Future research should explore what ecological trade-offs are being made in the process of protecting flagship species, what gaps exist in our knowledge of species under the protective umbrellas of these charismatic megafauna, and what alternative frames exist for communicating conservation through mass media.

Note

1 Publications in the study: *Boston Globe*; *Chicago Tribune*; *Discover* magazine; *Los Angeles Times*; *National Geographic*; *Nature* magazine; *New York Daily News*; *New York Times*; *New York Post*; *Popular Science* magazine; *San Francisco Chronicle*; *Science* magazine; *Scientific American*; *Smithsonian* magazine; *Economist*; *Time* magazine; *USA Today*; *Wall Street Journal*; *Washington Post*.

References

Albert, C., Luque, G. M., & Courchamp, F. (2018). The twenty most charismatic species. *PLoS One*, *13*(7), e0199149. doi:10.1371/journal.pone.0199149

Born, D. (2019). Bearing witness? Polar bears as icons for climate change communication in National Geographic. *Environmental Communication*, *13*(5), 649–663.

Bowes, C., & Gee, A. (2013). The first panda to leave China. BBC News. Retrieved from www.bbc.com/news/magazine-25136118

Boykoff, M. T. (2008). The cultural politics of climate change discourse in UK tabloids. *Political Geography*, *27*(5), 549–569.

Buchen, L. (2008). Are efforts to save the panda a giant waste of money? *Discover Magazine*. Retrieved from www.discovermagazine.com/planet-earth/are-efforts-to-save-the-panda-a-giant-waste-of-money

Buckingham, K. C., David, J. N. W., & Jepson, P. (2013). Environmental reviews and case studies: Diplomats and refugees: Panda diplomacy, soft "cuddly" power, and the new trajectory in panda conservation. *Environmental Practice*, *15*(3), 262–270.

Clucas, B., McHugh, K., & Caro, T. (2008). Flagship species on covers of US conservation and nature magazines. *Biodiversity and Conservation*, *17*(6), 1517.

Croke, V. C. (2006). *The lady and the panda: The true adventures of the first American explorer to bring back China's most exotic animal*. New York: Random House.

DeMarco, E. (2015). Save a panda, save the forest? *Science | AAAS*. Retrieved from www.sciencemag.org/news/2015/09/save-panda-save-forest

Di Minin, E., & Moilanen, A. (2014). Improving the surrogacy effectiveness of charismatic megafauna with well-surveyed taxonomic groups and habitat types. *Journal of Applied Ecology*, *51*(2), 281–288. doi:10.1111/1365-2664.12203

Dunaway, F. (2009). Seeing global warming: Contemporary art and the fate of the planet. *Environmental History*, *14*(1), 9–31.

Flather, C. H., Knowles, M. S., & Kendall, I. A. (1998). Threatened and endangered species geography. *BioScience*, *48*(5), 365–376.

Forbes. (2018). Pandanomics: What is giant panda conservation worth? Billions every year. *Forbes*. Retrieved from www.forbes.com/sites/grrlscientist/2018/06/29/pandanomics-what-is-giant-panda-conservation-worth-billions-every-year/

Glaser, B. G. (1965). The constant comparative method of qualitative analysis. *Social Problems*, *12*(4), 436–445.

Glaser, B. G., & Strauss, A. L. (2017). *Discovery of grounded theory: Strategies for qualitative research*. New York: Routledge.

Grossman, R. (2017). Love pandas? Thank the Roosevelts and the Field Museum. *Chicago Tribune*. Retrieved from www.chicagotribune.com/opinion/commentary/ct-teddy-roosevelt-panda-expedition-flashback-perspec-0604-20170603-story.html

Gusset, M., Maddock, A. H., Gunther, G. J., Szykman, M., Slotow, R., Walters, M., & Somers, M. J. (2008). Conflicting human interests over the re-introduction of endangered wild dogs in South Africa. *Biodiversity and Conservation*, *17*(1), 83–101.

Hamdani, Y. (2016, August 24). Undoing the discrimination against non-Muslims. *Daily Times Pakistan.*

Harris, J. B. C., Reid, J. L., Scheffers, B. R., Wanger, T. C., Sodhi, N. S., Fordham, D. A., & Brook, B. W. (2012). Conserving imperiled species: A comparison of the IUCN Red List and US Endangered Species Act. *Conservation Letters, 5*(1), 64–72.

Held, G. (2005). Magazine covers – a multimodal pretext-genre. *Folia Linguistica, 39*(1–2), 173–196.

Holland, J. (2019). China's new panda park will be three times bigger than Yellowstone. *National Geographic.* Retrieved from www.nationalgeographic.com/animals/2019/05/giant-pandas-national-park-china/

Huang, Z. A., & Wang, R. (2020). "Panda engagement" in China's digital public diplomacy. *Asian Journal of Communication,* 1–23.

Jacobson, S. K., Langin, C., Carlton, J. S., & Kaid, L. L. (2012). Content analysis of newspaper coverage of the Florida panther. *Conservation Biology, 26*(1), 171–179.

Jacobson, S. K., Morales, N. A., Chen, B., Soodeen, R., Moulton, M. P., & Jain, E. (2019). Love or loss: Effective message framing to promote environmental conservation. *Applied Environmental Education & Communication, 18*(3), 252–265.

Jepson, P., & Barua, M. (2015). A theory of flagship species action. *Conservation and Society, 13*(1), 95–104.

Kress, G. R. (2013). Multimodal discourse analysis. In *The Routledge handbook of discourse analysis* (pp. 61–76). New York: Routledge.

Kress, G. R., & Van Leeuwen, T. (1996). *Reading images: The grammar of visual design.* New York: Routledge.

Li, B. V. (2020). Conservation: Guarding panda land. *Current Biology, 30*(7), R302–R304.

Li, B. V., & Pimm, S. L. (2016). China's endemic vertebrates sheltering under the protective umbrella of the giant panda. *Conservation Biology, 30*(2), 329–339.

Lousley, C. (2016). Charismatic life: Spectacular biodiversity and biophilic life writing. *Environmental Communication, 10*(6), 704–718.

Manzo, K. (2010). Beyond polar bears? Re-envisioning climate change. *Meteorological Applications, 17*(2), 196–208.

Maxwell, M. (2004). *Setting the agenda: The mass media and public opinion.* Cambridge: Polity Press.

McCombs, M. E., & Shaw, D. L. (1972). The agenda-setting function of mass media. *Public Opinion Quarterly, 36*(2), 176–187.

McGowan, J., Beaumont, L. J., Smith, R. J., Chauvenet, A. L., Harcourt, R., Atkinson, S. C., . . . Beattie, A. (2020). Conservation prioritization can resolve the flagship species conundrum. *Nature Communications, 11*(1), 1–7.

Nicholls, H. (2011). The art of conservation. *Nature, 472*(7343), 287–289. doi:10.1038/472287a

Nicholls, H. (2012). *The way of the panda: The curious history of China's political animal.* New York: Pegasus Books.

Pimm, S. L., Jenkins, C. N., & Li, B. V. (2018). How to protect half of Earth to ensure it protects sufficient biodiversity. *Science Advances, 4*(8), eaat2616. doi:10.1126/sciadv.aat2616

Ren, G., Young, S. S., Wang, L., Wang, W., Long, Y., Wu, R., . . . Yu, D. W. (2015). Effectiveness of China's national forest protection program and nature reserves. *Conservation Biology, 29*(5), 1368–1377.

Slocum, R. (2004). Polar bears and energy-efficient lightbulbs: Strategies to bring climate change home. *Environment and Planning D: Society and Space, 22*(3), 413–438.

Swaisgood, R., Wang, D., &Wei, F. (2016). Ailuropoda melanoleuca (errata version published in 2017). *The IUCN Red List of Threatened Species*, e-T712A102080907.

Swaisgood, R. R., Wang, D., & Wei, F. (2018). Panda downlisted but not out of the woods. *Conservation Letters, 11*(1), e12355. doi:10.1111/conl.12355

Treves, A., Wallace, R. B., & White, S. (2009). Participatory planning of interventions to mitigate human – wildlife conflicts. *Conservation Biology, 23*(6), 1577–1587.

Walpole, M. J., & Leader-Williams, N. (2002). Tourism and flagship species in conservation. *Biodiversity and Conservation, 11*(3), 543–547.

Wei, F., Costanza, R., Dai, Q., Stoeckl, N., Gu, X., Farber, S., . . . Zhang, W. (2018). The value of ecosystem services from giant panda reserves. *Current Biology, 28*(13), 2174–2180, e7. doi:10.1016/j.cub.2018.05.046

10 Polar bears as cultural symbols

Threatening monsters and threatened species

Dorothea Born

In November 2019, a polar bear cub was born in Vienna's famous Schönbrunn Zoo. Before showing the cub to the public, the zoo announced a naming competition where people could suggest names for the newborn. This was largely propagated through the zoo's Instagram channel (@zooviennaschonbrunn) showing funny pictures and videos of the cute little baby bear. The cub's name – Finja – was finally revealed on International Polar Bear Day, February 27, 2020, an event organized by Polar Bears International to spread awareness about the dangers facing polar bears (Lam & Tegelberg, 2020). It was followed by a weekend full of informational campaigns about polar bear protection and climate change.

The images of Finja on the zoo's social media account follow a well-worn visual style. We see the baby bear "playing" in the mud, while captions tell us that "mother" Nora is "teaching" her new things. These images and captions exhibit an anthropomorphizing of polar bears, inviting viewers to empathize and identify with them. The zoo's social media campaign clearly builds on these connotations to attract new visitors. And by announcing the cub's name on International Polar Bear Day, the zoo demonstrated the intrinsic association of polar bears with climate change.

This hardly seems surprising since the polar bear (*Ursus maritimus*) has been the "poster child of climate change" (Hulme, 2009, p. 238) for quite some time. Back in 2006, an animated polar bear nearly drowned in former US Vice President Al Gore's Nobel Prize–winning film "An Inconvenient Truth." In the same year, a picture of a real polar bear made it onto the cover of *Time* magazine, titled "Be worried, be *very* worried," referring to climate change. In 2011, the World Wildlife Fund and Coca-Cola paired up for the fundraising campaign "Arctic Home," with the well-known Coca-Cola bears raising climate change awareness (Dunaway, 2009). In 2012, Greenpeace published a heart-breaking video of a polar bear stranded in London, sadly roaming the streets, with a depressing Radiohead song playing in the background (Tollmann, 2014). And in 2013, Greenpeace built an oversized polar bear marionette called Aurora that accompanied activists on the streets (Huggan, 2016). As these incidents demonstrate, many NGOs and media have used polar bears to illustrate or raise awareness of climate change.

Statement of the problem

Researchers in climate change communication have questioned the effectiveness of such images. For instance, O'Neill (2019a) showed how polar bear images have even been associated with climate change skepticism. Since "the power of the polar bear icon . . . rests on its emotional appeal" (Hulme, 2009, p. 242), it has been argued that the image's impact has "more to do with affect than cognition" (Manzo, 2010, p. 197). Images of seemingly helpless, drowning bears create negative reactions such as fear and hopelessness. Such images, on the one hand, raise awareness of climate change, but simultaneously may create the impression that climate change is a spatially and temporally remote issue (O'Neill & Nicholson-Cole, 2009). Further, polar bear images may even "block perceptual understanding of the humanistic dimension of the crisis" (Dunaway, 2009, p. 11). And Lam and Tegelberg (2020) argue that the brightness of typical images of the Arctic "anaestheticizes viewers to the bears' suffering" (p. 124). The criticism of such images has gone so far that George Marshall, the founder of climateoutreach.com, proclaimed: "Oh, not another bloody polar bear" (Marshall, 2007).

Linguist Brigitte Nerlich's blog post discusses whether polar bears are a "symbol," a "metaphor," a "visual metonym," or a "visual synecdoche" (Nerlich, 2019). And human geographer Saffron O'Neill (2019a, 2019b) nicely shows how these linguistic concepts can be used to theorize polar bear and climate change imagery. Notwithstanding such academic debates, polar bears today seem as famous as ever in climate change campaigns, as Finja's and other examples show (O'Neill, 2019b). They remain an icon for climate change because climate change itself is notoriously difficult to visualize (Manzo, 2010), and its worst large-scale effects lie in the future (Doyle, 2009). Using the polar bear icon gives climate change a face and triggers immediate emotional reactions (Born, 2019; Slocum, 2004), even though these reactions differ among viewers and may also be contradictory (O'Neill, 2019b).

While questions remain as to whether polar bears and their images should be used in climate change campaigns, as well as the effects of such use, we do know that changes in climate indeed endanger them (Owen & Swaisgood, 2008). A recent study predicts that they will likely be nearly extinct by the end of the century due to losses in the extent of sea ice (Molnár et al., 2020).

This chapter investigates how the polar bear gained this connection to anthropogenic climate change as "environmental images do not acquire iconic or representative status by themselves" (Hansen & Machin, 2013, p. 156). To understand the symbolic meaning of the polar bear icon, its effectiveness, its larger connotative meanings and cultural embeddedness, and the emotional responses it might trigger, this study looks at the larger visual history of polar bear images.

How did the polar bear get on the front page?

While polar bears now are perceived as vulnerable and threatened (Mooallem, 2013), they were portrayed mostly as fierce beasts threatening human life in early modern history (Engelhard, 2017). As recently as 1985, the documentary "Polar Bear Alert" on National Geographic TV depicted them as bloodthirsty monsters. Archibald (2015) argues that this change "from fierce to adorable" (p. 266) resulted from scientific advances in polar bear population monitoring and popular accounts of this research. This shift established them as icons for climate change. Hansen and Machin (2013) argue that for particular images to gain iconic status, it is necessary to have an ongoing process, drawing on "an extensive collection of semiotic resources" (p. 157).

This chapter investigates the "process of iconization" (Born, 2019, p. 650), exploring how polar bears became connected to the issue of anthropogenic climate change, and it compares the visual styles and deployments of polar bear images within two important popular science magazines, *National Geographic* and *GEO*.

Popular science magazines sit at the threshold between the mass media and scientific journals. They can be characterized as "hybrid spaces of science communication" (Born, 2020b, p. 3), where scientific truth claims are selected, recontextualized, narrated, and visualized in specific ways and thus transformed to create the distinct genre of popular-scientific knowledge. Popular science stories follow specific rules and norms of claim-making based on a high commitment to "scientificity" and on visual storytelling (Born, 2020b), thus fostering a "visual science literacy" (Bucchi & Saracino, 2016, p. 1) within modern societies.

Magazines build a community of readers who identify with the editorial strategies and values they promote (Abrahamson, 2007). These reader communities are important since popular science magazines are predominately read by well-educated people, who are often scientists or educators themselves, and thus have a multiplier function in society at large (Born, 2020b). These magazines not only educate people, but create powerful visual narratives about nature, science, and society. They visually construct climate change as an issue and employ specific narratives and images, such as the polar bear icon.

Comparing the US *National Geographic* and the German *GEO*, this chapter investigates the "cultures of viewing" (Hansen & Machin, 2013, p. 153) and uses of the polar bear icon in two countries with dramatically different climate change policies (Arlt, Hoppe, & Wolling, 2011) and communication strategies (Grundmann, 2007). *National Geographic*, founded in 1888, is known for its distinct visual style and has been called a "generator of icons" (Hawkins, 2010, p. 1). Originally founded as a scientific journal, advancements in photography around the beginning of the 20th century fostered the magazine's development into a more "popular" journal as the growing number and size of images came at the expense of more detailed scientific descriptions (Lutz & Collins, 1993). *GEO* was founded in Hamburg in 1979 by photojournalist Rolf Gillhausen, who at the time worked for *stern*, another German magazine. The editorial office of *stern* had, over time, acquired a large number of beautiful photographs but lacked the space to publish them all. *GEO*'s founding narrative says that *GEO* was created as a place to publish these images, which intrinsically connects the magazine to the visual (Born, 2020b).

Method

This chapter builds on previous research on the visual construction and communication of climate change in *GEO* and *National Geographic*, which analyzed an exhaustive sample of all feature articles and accompanying visuals about climate change published in these magazines over a 20-year time span from 1992 to 2012 (Born, 2020c). The author identified these articles using the *Complete National Geographic DVDs* and *GEO*'s internal archive. The articles focus on aspects of climate change, such as climate change science, politics, or mitigation, and some also focus on the consequences of climate change for polar bears. To investigate when and how the connection between climate change and polar bears was established, for this chapter's research the author included additional articles using the search term "polar bear." Thus, some articles that did not primarily deal with climate change but focused on polar bear biology and behavior are included in the sample. See Table 10.1 for a list of articles selected for the chapter's research question.

The first difference between *GEO* and *National Geographic* was their number of articles and images connected to polar bears. *National Geographic* published seven articles on polar bears, or polar bears and climate change, leading to a sample of 72 images. *GEO* published three articles and 63 images during the same period. These numbers need to be put in relation to the magazines' total number of climate change articles. Four of the seven *National Geographic* articles were also part of the sample on climate change, which consisted of 30 articles. The three *GEO* articles were all part of the sample on climate change, which, however, was larger overall, containing 50 articles. This suggests that polar bears were more prominent in *National Geographic*'s climate change coverage than in *GEO*'s.

The overall methodological approach for this study is multimodal critical discourse analysis (see Machin & Mayr, 2012). This approach builds on the assumption that different discursive modes work together to create meaning. Thereby, multimodal discourses are social semiotic practices, which both shape and are shaped by society (Machin & Mayr, 2012).

To understand the interplay between the verbal and visual, the author first removed all textual elements from the images using Photoshop. The images were then analyzed in isolation before analyzing them together with the accompanying texts and captions. The images were grouped using type-building (Müller-Doohm, 1996). The selected prototype images were further analyzed using a combination of semiotics and iconography (see Born, 2019).

Findings

National Geographic*: a process of iconization*

During this period, *National Geographic* appeared to present polar bears as increasingly connected to the issue of anthropogenic climate change in a

Table 10.1 List of articles on polar bears and polar bears and climate change published in *GEO* and *National Geographic* between 1992 and 2012.*

Magazine	*Title and Subtitle of Articles*	*Author*	*Photographer*	*Month and Year Published*
National Geographic	*Polar Bears*. Stalkers of the High Arctic	John L. Eliot	Flip Nicklin (+ 1 photograph by Norbert Rosing)	1/1998
GEO	*Eiszeit ist ihr wahres Leben. Die Nomande von der Hudson Bay. [Living in the Ice Age. The Nomads from Hudson Bay]*	Lothar Frenz	Norbert Rosing	2/1999
National Geographic	*Bear beginnings*. New Life on the Ice	Norbert Rosing	Norbert Rosing	12/2000
National Geographic	*White on White*. Polar Bears.	John L. Eliot	Norbert Rosing	2/2004
National Geographic	*Refuge in White*. Winter in a Canadian National Park	John L. Eliot	Norbert Rosing	12/2005
GEO	*Wärembilder aus dem Eis. [Thermal images from the ice.]*	Christine Heidemann	Thorsten Milse	2/2006
National Geographic	*Life at the Edge*. On the frontier of a frozen ocean, rising temperatures imperil wildlife whose survival depends on ice.	Paul Nicklen	Paul Nicklen	6/2007
GEO	*Jahre im Eis. [Years in the Ice]*	Hania Luczak	Mireille de la Lez	11/2007
National Geographic	*Ice Paradise*. The rich life of Svalbard, Norway's Arctic archipelago, faces a creeping thaw.	Bruce Barcott	Paul Nicklen	4/2009
National Geographic	*On Thin Ice*. The Arctic is warming so fast that by 2050 it may be largely ice free in summer. Without their frozen hunting platform, how will polar bears survive?	Susan McGrath	Florian Schulz	7/2011

* Table courtesy of Dorothea Born.

distinct "process of iconization" (Born, 2019, p. 650). This process had three distinct phases, each in a different visual style.

In the three articles published during the first phase, 1998–2004, polar bears were often presented in an anthropomorphized way. Most images were close-up shots, displaying them in poses that could easily be interpreted as human-like behavior. These pictures were supported by captions stating that the bears were "play-fighting" or "cuddling," while females were described as protecting but also occasionally punishing their cubs (Born, 2019).

For example, Norbert Rosing's 1998 *National Geographic* photograph titled "Polar Bear at Ease" (Figure 10.1) shows a bear lying on the ice, much as a child would. Its sagging jowls seem to make it look calm but a bit sad. While today one might associate this image with climate change, articles during this first phase focused primarily on polar bears' behavior and biology. "Climate change" was not part of these articles' vocabulary. Occasionally, however, article authors mentioned "warming trends" or earlier spring ice breakup.

This is in accordance with the author's previous research, which showed that *National Geographic* articles during the late 1990s and early 2000s left the question of the causes of climate change unanswered (see Born, 2020a). Articles in 2004 finally claimed a connection existed between human burning of fossil fuels and climate change. The magazine achieved this through two parallel discursive strategies. On the one hand, it framed climate change as more and

Figure 10.1 Anthropomorphized bear resting on the ice in January 1998.

Source: Photograph by Norbert Rosing, published with permission from *National Geographic Creative*.

more scientifically certain and, on the other hand, as more and more unnatural and, therefore, caused by humans.

Once anthropogenic climate change was established as a "public fact" in 2004 (see Born, 2020a, p. 110), the visual discourse on polar bears shifted. In the second phase, which comprises one article published in 2005, they became verbally and visually connected to their Arctic environment; and the Arctic ecosystem itself was the subject of articles presenting the landscape and wildlife. For example, Rosing's *National Geographic* photograph from 2005 (Figure 10.2) shows a mother cuddling her two cubs in the snow. Still anthropomorphic, the family theme predominates, but now the sow appears to be sheltering her cubs from the Arctic cold.

This transitional phase, from 2005 to 2006, was followed by a third phase from 2007 to 2012, in which the magazine published three articles depicting polar bears as being in danger. By that time, the iconic visuality of polar bear images was fully developed. A 2007 *National Geographic* photograph by Paul Nicklen (Figure 10.3) shows a wide shot of two bears on a sunny day with their necks stretched up to the sky, floating on a tiny iceberg in the middle of a deep blue sea. The caption begins, "A mother polar bear and her two-year-old cub stranded."

The close-up shots from the first phase established polar bears as individuals, and the "the image of a mother bear with cubs resonates with . . . the figure of the ideal victim . . . as a weak, morally blameless female" (Lam & Tegelberg,

Figure 10.2 Female bear sheltering her two cubs in December 2005.

Source: Photograph by Norbert Rosing, published with permission from *National Geographic Creative*.

Figure 10.3 Two polar bears stranded on an iceberg in June 2007.

Source: Photograph by Paul Nicklen, published with permission from *National Geographic Creative*.

2020, p. 122 ff.). In contrast, the long shots of the third phase make the bears appear more generic and, thus, more symbolic (Van Leeuwen, 2000). Articles in the third phase are clearly connected to anthropogenic climate change, with titles such as "Life at the Edge" (June 2007, *National Geographic*) and "On Thin Ice" (July 2011, *National Geographic*), and describe as well as portray the bears as endangered by global climate change.

All three phases contributed to establishing polar bears as icons for climate change. The anthropomorphized depictions of the first phase established them as figures of identification for readers, meaning that readers identified with them. The second phase connected them to their Arctic environment, which, in the last phase, established them as symbols for an ecosystem endangered by a warming climate.

Polar bears in GEO*: nature "red in tooth and claw"*

Examination of the German popular science magazine *GEO* did not disclose a continuous and changing visual discourse on polar bears, unlike *National Geographic*. A major reason is because only three articles on polar bears appeared in the analysis period. These three articles (in 1999, 2006, and 2007) exhibited a visual language similar to the anthropomorphized portrayals in *National Geographic*'s first phase. Most polar bear images in *GEO* were close-up shots, and in many pictures the bears appeared human-like (e.g., through mommy–baby interactions), seemingly expressing emotions.

The principal difference in coverage between *GEO* and *National Geographic* was that as far back as 1999, *GEO* connected polar bears to climate change. This also relates to the different narratives the two magazines constructed about anthropogenic climate change. *National Geographic* framed it as a question of scientific proof and uncertainty and thus first presented anthropogenic climate change as a scientific and public fact in 2004, once a broad scientific consensus was established (Born, 2020a). *GEO*, on the other hand, framed it as a political and moral concern, as well as a scientific fact. In light of Germany's political decision-making, questions of scientific certainty faded into the background, which *enabled GEO* to present climate change as factual from early on (Born, 2020c).

Another difference became apparent in *GEO*'s verbal discourse. Articles from 1999 and 2007 focus on the photographers' perspective, recounting their experiences, motivations, and difficulties while photographing in the Arctic. They also recount the photographers' dangerous encounters with polar bears. In contrast to the imagery, which is similar to the cuddly, anthropomorphized bears in *National Geographic*'s first phase, these verbal accounts thus portray polar bears as dangerous predators. A 2007 article titled "Years in the Ice" exemplifies this different characterization of polar bears, even though not directly depicting them.

On one page is a picture of a newborn ringed seal, which appears cute but also helpless, and immediately triggers protective instincts. The picture on the next page shows the fate of the seal: Readers see a trail of blood on the white snow, traces of a polar bear's last meal.

The accompanying text describes how the photographer first watched the seal pup's birth and then how the seal cow got stuck in her breathing hole. While the photographer listens to the desperate cries of the seals, she does not interfere as she wants to document, but not interfere with, nature. Visiting the site of the drama the next day, she finds only traces of blood. This contrasts with an article in the same year in *National Geographic* called "Life at the Edge," where a picture of a seal pup is accompanied by a caption stating that it had lost its way on the ice but that the photographer pitied the helpless creature and carried it to the edge of the water.

The two magazines evoked contrasting imaginings of nature. Through its aesthetically appealing photography, *National Geographic* created an image of nature as a beautiful, pristine wilderness in need of human protection (Cosgrove, 2008). *National Geographic* images evoked feelings of awe as well as pity, with few references to disturb this romanticized vision of nature. In contrast, *GEO* evoked a nature "red in tooth and claw" (Cook, 2007, p. 49), where one can witness the Darwinian struggle for existence. *GEO* depicted and described nature as both terrible and beautiful. In *GEO*, the Arctic wilderness is not romanticized, but rather presented as a terrifying place where only the toughest survive. Polar bear images in the two magazines are also differently gendered.

While the female polar bears in *National Geographic* are the "ideal victims" needing protection, the polar bear and the Arctic landscape in *GEO* appear as "hypermasculine opponents" of humans (Lam & Tegelberg, 2020, p. 125).

Discussion

There were important differences between the two magazines in deploying polar bears as icons for climate change. First, polar bears were more prominent in *National Geographic's* overall climate change discourse, as seen by the proportion of articles dealing with climate change and polar bears to the total number of climate change articles. Second, the two magazines established and stabilized climate change differently, which also led to different uses of the polar bear icon. In *National Geographic*, climate change was gradually established as a scientific and public fact. Similarly, polar bears became icons for climate change over a longer, continuous visual discourse. *GEO* always treated anthropogenic climate change as factual and always connected it with polar bears. Thus, in *GEO*, there was no *process of iconization* through a continuous visual discourse. Third, this also becomes apparent in the different visual styles they deployed when portraying polar bears. While the visual language changes in *National Geographic* from anthropomorphized, close-up shots to long shots connecting polar bears to their Arctic environment and presenting them as endangered by climate change, the visual language in *GEO* did not change over time. Pictures of polar bears in *GEO* are mostly close-up shots and similar to the anthropomorphized style in *National Geographic's* earlier articles, while the verbal accounts focusing on the photographers' perspectives also describe polar bears as dangerous predators.

Teddy bear tales

The visuality of the "anthropomorphized phase" within *National Geographic* parallels the highly anthropomorphized animated polar bears happily drinking Coca-Cola, a US brand famous around the world. This anthropomorphizing of bears, in turn, is based on the iconology of the teddy bear, which has its roots more than 100 years ago. In 1902, so the story goes, US President Theodore Roosevelt was on a bear hunt when his fellow hunters captured a bear for him to shoot. Instead of killing it, he pitied the bound and helpless creature. A newspaper cartoonist took up this tale, and soon after, the stuffed animal was born, named after President "Teddy" Roosevelt (Bieder, 2005). In 1903, the Steiff plush toy company in Germany received a 3,000-unit American order for Teddy Bears in response to the "Teddy" Roosevelt bear cartoon (Ebling, 1997). And so, the Teddy Bear craze and the love for bears was born in both the United States and Germany.

In 2006, the Center for Biological Diversity petitioned the US government to put polar bears on the list of endangered species, and in 2008 they were listed as "threatened" under the Endangered Species Act (ESA) (Owen & Swaisgood,

2008). The rationale was that if polar bears indeed merited the status of endangered species due to climate change, the George W. Bush administration would have to acknowledge climate change as a threat (Mooallem, 2013). This strategy did not fully work, as the ESA listing explicitly excludes climate change mitigation activities. However, it generated a lot of media attention around polar bears (Owen & Swaisgood, 2008). The process of iconization in *National Geographic* thus reflects and also (re)produces and solidifies these larger cultural and national imaginations.

In Germany, polar bears and climate change were connected, as exemplified by the polar bear Knut, who was born in 2006 in the Berlin Zoo, just when the climate icon narrative was taking hold in the United States. Rejected by his mother and subsequently brought up on the bottle by his zookeeper, Knut attracted many visitors, as well as massive media attention. This so-called "Knutmania" (Engelhard, 2017, p. 25) not only earned the zoo a fair amount of money but also supported many climate change campaigns. For example, German environmental minister Sigmar Gabriel symbolically adopted Knut and used the bear's pictures for his climate change awareness campaigns. Leonardo DiCaprio, a US actor and climate change activist, was photoshopped with Knut on the cover of *Vanity Fair* (Engelhard, 2017).

This shows the different political and cultural context for using the polar bear icon in Germany and the United States because deploying polar bear images in Germany also meant a reference to Knut. Yet, an image of Knut and an image of any lonely polar bear on an iceberg have different iconic functions, as the next section explains.

Different genres of icons

According to Perlmutter (2006), we need to differentiate between "discrete" and "generic" icons (p. 54). While *discrete icons* are single images of specific events that happened in history, such as pictures of the famous "Tank Man" at Tiananmen Square in Beijing, *generic icons* are repeatedly shown to describe or represent specific phenomena (Perlmutter, 2006). The icon of the polar bear falls into the latter category. The image of a polar bear on an ice floe has become so generic that we immediately associate it with climate change. It is this type of generic icon that was established and reproduced through images in *National Geographic*. In contrast, images of the concrete polar bear Knut or other polar bears raised in zoos, such as Flocke in Nuremberg (Passoth, 2010) and Finja in Schönbrunn, show a distinct polar bear, but are also meant to stand for something larger than just what they depict.

Different climate change discourses

The different genres of icons, in turn, may reflect the alternate political discourses around climate change in Germany and the United States. While skeptical voices regarding anthropogenic climate change are not unusual in the

United States (Dunlap, 2013), climate change skepticism in Germany is relatively low among citizens, as well as in the media (Schäfer, 2016).

In the United States, where climate change is still controversial, the generic polar bear icon is needed to raise climate change awareness. As such images trigger emotional reactions, they help to "emotionally anchor" (Höijer, 2010, p. 722) the issue. *National Geographic* used this function as one strategy to create awareness of anthropogenic climate change in a highly politicized, controversial, and still ongoing public debate over the existence of (anthropogenic) climate change that peaked in 1997 and was fueled by conservative think tanks and right-wing politicians (Grundmann, 2007; Dunlap, 2013). Furthermore, since polar bears are native to the United States, the generic icon speaks well to a romanticized idea of wilderness that is part of the American identity (Cosgrove, 2008).

In both countries, polar bears in zoos sometimes receive names. As distinct "personalities," they can then be used for political campaigns or to foster individual climate change engagement.

Conclusion

It becomes clear after investigating the visual history of polar bear images in varying contexts that the polar bear did not become an icon for climate change by chance. Rather, it was built up through a process of iconization, based on a set of semiotic resources and cultural associations (Hansen & Machin, 2013). The iconic function of polar bear images is highly context dependent and does not work in every setting.

Even though anthropogenic climate change indeed imperils polar bears (Owen & Swaisgood, 2008), they are not the only Arctic species adversely affected. Focusing on polar bears as "charismatic megafauna" and building on their "nonhuman charisma" (Lorimer, 2007, p. 912) seem to imply that only what is cute and is a figure of identification is worthy of protection.

The *generic icon* of the polar bear as a projection surface humans can identify with also builds on the colonial idea of the Arctic as a vast, untouched, and uninhabited space, waiting to be conquered by Western explorers (Westerstahl-Stenport & Vachula, 2017). The polar bear icon may make invisible other Arctic species and even the native peoples of the Arctic, though they are also affected by a changing climate (Engelhard, 2017). The negative impacts of polar bear hunting restrictions on native peoples may also be made invisible by concerns for the cuddly polar bear and climate change. Furthermore, depicting polar bears within the Arctic landscape, surrounded by ice and snow, using cool and calm colors, does not show human influences on these landscapes and renders the suffering of polar bears distant (Lam & Tegelberg, 2020).

Discrete polar bear icons are either zoo inhabitants, such as Knut and Finja, or the stars of Disney-style nature documentaries, such as the family of polar bears featured in the film *Earth* (2007). The former carry little reference to their "natural" living conditions, while the latter tell a story set in the Arctic.

Behind both the generic and distinct icons lie commercialization and economic exploitation. While no longer hunted with guns, they are now hunted with cameras to produce the "perfect" photograph or to tell an idealized story for documentary film audiences. And marketing campaigns with cute cubs on online platforms such as Instagram increase follower numbers, attract new visitors, and sell polar bear merchandise.

The use of polar bear images to raise climate change awareness may actually inhibit a broader cognitive understanding of the issue (Manzo, 2010). The icon may attract attention and trigger emotional reactions, but it may not make the larger causes of nature exploitation and climate change more visible (Hansen & Machin, 2013).

Climate change visuals, such as polar bear icons, are generally connected to certain narratives and discourses. To improve the iconography of climate change, new climate change narratives should be found that do not focus only on charismatic animals, but also on the larger economic causes and circumstances of climate change. These narratives should not depict humans and nature as separated, but as intertwined, co-dependent, and equally affected by climate change.

Acknowledgments

This article builds on research previously published in Born, D. (2017). Bearing witness? Polar bears as icons for climate change communication in *National Geographic. Environmental Communication, 13*(5), 649–663, and in Born, D. (2017). Sind wir nicht alle Eisbären? [Aren't we all polar bears?]. *Sans Phrases – Zeitschrift für Ideologiekritik, 10*, 86–75.

The author is grateful to the staff of *National Geographic* and *GEO* for their contributions.

References

Abrahamson, D. (2007). Magazine exceptionalism. *Journalism Studies, 8*(4), 667–670.

Archibald, K. (2015). From fierce to adorable: Representations of polar bears in the popular imagination. *American Review of Canadian Studies, 45*(3), 266–282.

Arlt, D., Hoppe, I., & Wolling, J. (2011). Climate change and media usage: Effects on problem awareness and behavioural intentions. *International Communication Gazette, 73*(1–2), 45–63.

Bieder, R. E. (2005). *Bear*. London: Reaktion Books.

Born, D. (2017). Sind wir nicht alle Eisbären? [Aren't we all polar bears?]. *Sans Phrases – Zeitschrift für Ideologiekritik, 10*, 86–75.

Born, D. (2019). Bearing witness? Polar bears as icons for climate change communication in *National Geographic. Environmental Communication, 13*(5), 649–663.

Born, D. (2020a). Visual climate communication: Making facts and concerns in popular science magazines. In U. Felt & S. Davies (Eds.), *Exploring science communication: A science and technology studies approach* (pp. 109–128). London: Sage.

Born, D. (2020b). Making science popular: Producing the German popular science magazine *GEO*. Manuscript submitted for publication.

Born, D. (2020c). *Images and imaginations of climate change: The construction of climate change in popular science magazines* (PhD thesis). University of Vienna, Vienna.

Bucchi, M., & Saracino, B. (2016). "Visual science literacy": Images and public understanding of science in the digital age. *Science Communication*, *38*(6), 812–819.

Cook, D. (2007). Nature, red in tooth and claw. *Continental Philosophy Review*, *40*(1), 49–72.

Cosgrove, D. (2008). Images and imagination in 20th-century environmentalism: From the Sierras to the poles. *Environment and Planning A*, *40*(8), 1862–1880.

Doyle, J. (2009). Seeing the climate? The problematic status of visual evidence in climate change campaigning. In S. I. Dobrin & S. Morey (Eds.), *Ecosee: Image, rhetoric, nature* (pp. 279–298). Albany, NY: SUNY Press.

Dunaway, F. (2009). Seeing global warming: Contemporary art and the fate of the planet. *Environmental History*, *14*(1), 9–31.

Dunlap, R. E. (2013). Climate change skepticism and denial: An introduction. *American Behavioral Scientist*, *57*(6), 691–698.

Ebling, A. (1997, December 29). Hot investment tip: Teddy bears. *Forbes*. Retrieved from www.forbes.com/forbes/1997/1229/6014062a.html#3a526ab82f50

Engelhard, M. (2017). *Ice bear: The cultural history of an Arctic icon*. Seattle, WA, and London: University of Washington Press.

Grundmann, R. (2007). Climate change and knowledge politics. *Environmental Politics*, *16*(3), 414–432.

Hansen, A., & Machin, D. (2013). Researching visual environmental communication. *Environmental Communication*, 7(2), 151–168.

Hawkins, S. L. (2010). *American iconographic: National Geographic, global culture, and the visual imagination*. Charlottesville, VA: University of Virginia Press.

Höijer, B. (2010). Emotional anchoring and objectification in the media reporting on climate change. *Public Understanding of Science*, *19*(6), 717–731.

Huggan, G. (2016). Never-ending stories, ending narratives: Polar bears, climate change populism, and the recent history of British nature documentary film. In J. Nyman & N. Schuurman (Eds.), *Affect, space and animals* (pp. 13–24). Abingdon, UK, and New York: Taylor and Francis.

Hulme, M. (2009). *Why we disagree about climate change: Understanding controversy, inaction and opportunity*. Cambridge, UK, and New York: Cambridge University Press.

Lam, A., & Tegelberg, M. (2020). #Sickbear: Photographing polar bears as ideal nonhuman victims. In *Criminal anthroposcenes: Media and crime in the vanishing Arctic* (pp. 107–143). Cham: Springer International Publishing.

Lorimer, J. (2007). Nonhuman charisma. *Environment and Planning D: Society and Space*, *25*(5), 911–932.

Lutz, C., & Collins, J. L. (1993). *Reading National Geographic*. Chicago and London: University of Chicago Press.

Machin, D., & Mayr, A. (2012). *How to do critical discourse analysis: A multimodal introduction*. Los Angeles: Sage.

Manzo, K. (2010). Beyond polar bears? Re-envisioning climate change. *Meteorological Applications*, *17*(2), 196–208.

Marshall, G. (2007). *Carbon Detox: Your step-by-step guide to getting real about climate change*. London: Hamlyn Gaia.

Molnár, P. K., Bitz, C. M., Holland, M. M., Kay, J. E., Penk, S. R., & Amstrup, S. C. (2020). Fasting season length sets temporal limits for global polar bear persistence. *Nature Climate Change*. doi:10.1038/s41558-020-0818-9

Mooallem, J. (2013). *Wild ones: A sometimes dismaying, weirdly reassuring story about looking at people looking at animals in America*. New York: Penguin Books.

Müller-Doohm, S. (1996). Die kulturelle Kodierung des Schlafens oder: Wovon das Schlafzimmer ein Zeichen ist. *Soziale Welt, 47*(1), 110–123.

Nerlich, B. (2019, April 19). It's an icon, it's a symbol: It's a polar bear!? (blog post). Retrieved from http://blogs.nottingham.ac.uk/makingsciencepublic/2019/04/19/its-an-icon-its-a-symbol-its-a-polar-bear/

O'Neill, S. J. (2019a). More than meets the eye: A longitudinal analysis of climate change imagery in the print media. *Climatic Change, 163*(1).

O'Neill, S. J. (2019b). Climate bear, visual metonym. Manuscript submitted for publication.

O'Neill, S. J., & Nicholson-Cole, S. (2009). "Fear won't do it": Promoting positive engagement with climate change through visual and iconic representations. *Science Communication, 30*(3), 355–379.

Owen, M. A., & Swaisgood, R. R. (2008). On thin ice: Climate change and the future of polar bears. *Biodiversity, 9*(3–4), 143–148.

Passoth, J. H. (2010). Diskurse, Eisbären, Eisberge: Material-Semiotische Verwicklungen und der Klimawandel. In M. Voss (Ed.), *Der Klimawandel: Sozialwissenschaftliche Perspektiven* (pp. 49–59). Wiesbaden, Germany: VS Verlag für Sozialwissenschaften.

Perlmutter, D. D. (2006). Hypericons: Famous news images in the internet-digital-satellite age. In P. Messaris & L. Humphreys (Eds.), *Digital media: Transformations in human communication* (pp. 51–64). New York: Peter Lang.

Schäfer, M. S. (2016). Climate change communication in Germany. In *Oxford research encyclopedia of climate science: Interactive factory*. Oxford, UK: Oxford University Press.

Slocum, R. (2004). Polar bears and energy-efficient lightbulbs: Strategies to bring climate change home. *Environment and Planning D: Society and Space, 22*(3), 413–438.

Tidmarsh, A. (Producer), Tasiolius, S. (Producer), Fothergill, A. (Director), & Linfield, M. (Director). (2007). *Earth* (Documentary film). London: BBC Worldwide.

Tollmann, V. (2014). The uncanny polar bear: Activists visually attack an overly emotionalized image clone. In B. Schneider & T. Nocke (Eds.), *Image politics of climate change: Visualizations, imaginations, documentations* (pp. 249–272). Bielefeld: Transcript.

Van Leeuwen, T. (2000). Semiotics and iconography. In T. Van Leeuwen & C. Jewitt (Eds.), *The handbook of visual analysis* (pp. 92–118). London: Sage.

Westerstahl-Stenport, A., & Vachula, R. S. (2017). Polar bears and ice: Cultural connotations of Arctic environments that contradict the science of climate change. *Media, Culture & Society, 39*(2), 282–295.

11 Societal attitudes towards hellbender salamander conservation

The roles of traditional and social media

Caleb R. Hickman and Shem D. Unger

Throughout the history of the United States, legislation has been enacted that led either to the extermination or the protection of many species and their habitats. Decisions to exterminate were often based on a variety of reasons related to financial loss, such as wolf predation on livestock (Musiani & Paquet, 2004); or financial gain, for example, the market fur trade (Dunaway, 1994); or reducing densities to impact others, such as the slaughter of American bison to remove a Native American food source (Dary, 1989). Protections have occurred for many reasons too, including financial gain, for example, by instituting wild game and fish license fees and taxes (Kallman, 1987); improving human health, by preventing the interstate movement of diseased animals (Smith et al., 2009); and promoting ecosystem stability, such as the ecosystem services provided by biodiversity (Portman, 2013). Shaw (1985) articulated this history of management by defining temporal eras in the United States that could be categorized, largely through legislation, to reflect the priorities of the day.

Shaw (1985) explained a timeline of events in US history that marked changes in the ways people managed natural resources. The five Management Eras defined by Shaw (1985) were Exploitation (1850–1899), Protection (1900–1929), Game Management (1930–1965), Environmental Management (1966–1979), and Conservation Biology (1980–present). These eras are based on important legislation that led to changes in how society manages natural resources. For example, the Era of Exploitation (also termed Overexploitation) began with the transformation of the North American continent by European settlers, especially after the Industrial Revolution. Society in the Era of Exploitation had largely continued to promote an agrarian perspective (nature needed to be tamed) where people lived by manifest destiny perspectives (God provided nature and land to the colonizers). This era also promoted manufacturing and technology that resulted in large changes in the landscape, such as the formation of cities, farms, and railroads, thereby altering the environment. These large changes resulted in the near-extinction of species that society wished to control or possess, such as the passenger pigeon (*Ectopistes migratorius* – now extinct), mountain lion (*Puma concolor*), and grizzly bear (*Ursus arctos horribilis*).

Many early writings about nature sensationalized the danger of some animals and the consumption of others, so it is no surprise that these same animals became over-exploited. This era also brought about the early exploration of nature that led to great discoveries about the natural world, and novel thoughts in conservation. Prominent among them was Charles Darwin's *The Origin of Species* (1859) that discussed, in some detail, the human role in the environment as part of nature, rather than as the dominant species.

After over a century since the Era of Exploitation, the Conservation Biology Era was marked by the development of a new interdisciplinary field that sought to preserve at-risk biological diversity (Soule, 1985). The Conservation Era was especially based on major legislation that built from the Environmental Era to fund actionable research and management to preserve species. For example, in 2005, the Wildlife Action Plan was implemented; all states and territories were required to develop comprehensive wildlife management strategies, including plans for conserving non-game species and biodiversity.

Legislation is often driven by the voting public's perceived needs, but where does the initial voice of the people come from? The public voice often comes from media, through printed newspaper articles, magazines, and even today's online formats, reflecting the attitudes of the time. Through media, people are interviewed for news pieces and researchers are consulted about their results, but these two are rarely compared. Readership has always been a major hallmark of a successful story, where the writer might leverage controversy or dramatic effects to increase appeal. Social media can gain public engagement more easily than traditional media but is often biased towards charismatic or eye-catching species like wolves, American bison, eagles, and bears (Roberge, 2014). However, social media are undergoing a fast-paced evolution and can provide a method for researchers to act as a public voice for science (Bik & Goldstein, 2013). But how would the public respond to less charismatic and less-easily observed species?

One such poorly recognized species is the hellbender salamander (*Cryptobranchus alleganiensis*). Hellbenders are fairly widespread (Petranka, 1998) and have a tumultuous history of conservation, persecution, endearment, and public fear (Mullendore et al., 2014; Perry-Hill et al., 2014; Reimer et al., 2014). Hellbenders rely on clear, cool, well-oxygenated streams (Hillis & Bellis, 1971) with ample larger rock shelters (Rossell et al., 2013), and primarily consume crayfish as adults (Netting, 1929). Anglers are often alarmed to see these large, fully aquatic salamanders with small beady eyes and a wrinkled appearance (Petranka, 1998; Phillips & Humphries, 2005) (Figure 11.1).

Because of their appearance, hellbenders have been misunderstood and vilified, with myths regarding them as either "poisonous" or "bad as consumers of all the fish in a stream" (Nickerson & Briggler, 2007). Historically, hellbenders were characterized as one of the dominant organisms in stream systems within their former geographic range (Nickerson & Mays, 1973). Sadly, within their geographic range, many populations face numerous threats resulting in dramatic declines (Burgmeier, Unger, Sutton, & Williams, 2011; Wheeler, Prosen,

Figure 11.1 A hellbender salamander, the largest amphibian species found in North America, photographed at Chattahoochee National Forest.

Source: Photograph by Jerry Whitten, published with permission.

Mathis, & Wilkinson, 2003). In addition, the Ozark hellbender subspecies (*C. a. bishopi*) is now listed under Section 7 of the Endangered Species Act (ESA) as federally endangered, partially because of its limited distribution, while the Eastern hellbender (*C. a. alleganiensis*) was reviewed but not recommended for listing status. It is unknown if public support could influence the decision by the US Fish & Wildlife Service (USFWS) not to list Eastern hellbenders. This species' decline, along with associated research, has brought about new attention to the plight of the hellbender in the news and social media.

Conservation efforts for at-risk species often require funds for research and management, as well as public support. To obtain public support, the news media may provide an ideal outlet for species facing declines, numerous threats, and potential extinction. Therefore, it is of the utmost importance that media and conservation align to inform the general public, and ultimately those who create environmental policy. That includes not only information about specific conservation practices but also emphasizing the need to disseminate accurate natural history information regarding a species' biology and general importance in the ecosystem.

Traditional media outlets, such as newspapers, have conveyed both accurate stories and false narratives regarding the potential harmfulness of many species often overlooked by society, such as snakes, alligators, sharks, and others typically thought of in a negative light (Fogelman, 1988; Pandey, Pandey, Devkota, & Goode, 2016). One example of biased perceptions resulting in consequences appeared when wolf populations were extirpated from portions of North America because they were depicted in books, magazines, and news reports as dangerous to people and the economy through their predation on livestock (Hamilton, Lambert, Lawhon, Salerno, & Hartter, 2020). The early American media perpetuated public fear of wolves, which caused policy makers to implement a strategy to systematically slaughter the animals. More recently, attacks by dogs are often mistakenly identified as possible wolf attacks, highlighting the persistence of these fears among the public today (Caniglia et al., 2016). To enact conservation policy and prevent the extinction of at-risk species like wolves, new forms of engaging the public are needed as part of best management practices.

Emerging social media, such as Facebook, Instagram, Snapchat, YouTube, and Twitter, provide a valuable tool to promote fact-based natural history, dispelling myths and misconceptions, as well as to engage and involve the public in citizen science. Public attitudes can shape the management of native species (Eeden, Dickman, Ritchie, & Newsome, 2017). Because more people than ever are using the internet to find information (Purcell, Brenner, & Rainie, 2012), scientists are urged to provide more direct links through online and social media platforms so empirical scientific research can guide public perception (Brossard & Scheufele, 2013). However, recent studies have shown only a small percentage of scholars are active on social media platforms, such as Twitter (Costas, Honk, & Franssen, 2017), and scientists rarely communicate via online media (Lo, 2016; Howell, Nepper, Brossard, Xenos, & Scheufele, 2019). Social media concomitantly has the potential to both influence the spread of accurate science and to misinform (Waszak, Kasprzycka-Waszak, & Kubanek, 2018), which can directly affect public perception and support for species conservation.

Statement of the problem

It is often difficult for conservationists and biologists to fully understand whether their work is impactful to society at large through changes in policy. This chapter attempts to determine whether the media can be leveraged to detect social attitudes towards a species of conservation concern. The authors created a conceptual framework to diagram the connections of how the media may be influenced by science and public perception, and also how the media may inform public perception (Figure 11.2). Through the filter of public perception, the media feeds into policy and decision-making. For example, newspaper articles can reflect public perceptions of hellbender salamanders as dangerous or threatening to fisheries. These negative perceptions can be associated with large-scale policy changes, leading to persecution. However, this conceptual framework shows how scientists can provide valid research and input about actual threats,

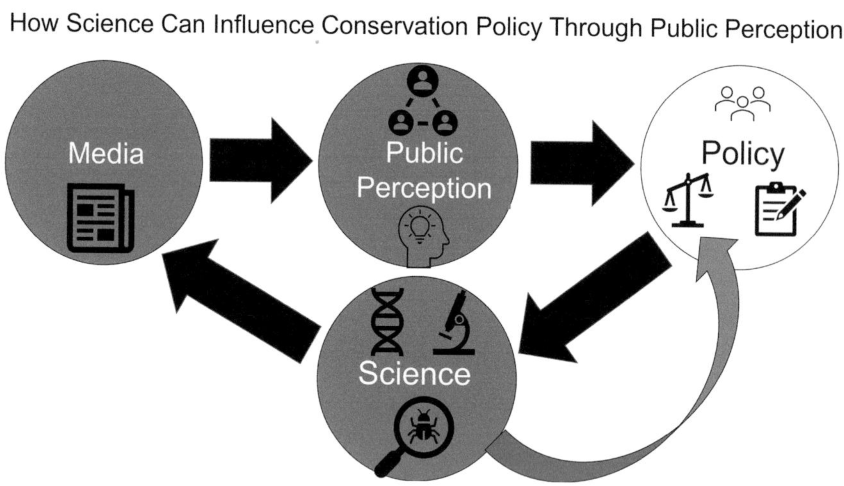

Figure 11.2 Conceptual diagram shows how scientists can influence policy through interaction with print and online media. Darker circles are where social information is developed and where traditional media can drive conservation policy through science-to-media process. After information about the state of the science is shared to media platforms, information is processed through the public filter and passed on to policy makers (light circle). Policy makers can then support conservation and further science through funding and changes in law.

Source: Image courtesy of Caleb R. Hickman and Shem D. Unger.

or lack thereof, from hellbenders. Scientists can provide information to journalists to develop fact-based news articles and can then help dispel myths and placate public perceptions. Therefore, this conceptual diagram, along with reviews of media sources, investigates whether promoting better scientific practices will aid in positive public perceptions and associated conservation policy decisions. Using a content analysis method, this study focused on media about hellbender salamanders that have suffered a history of human persecution and negative perceptions. In addition, hellbender salamanders have experienced dramatic population declines with a need for conservation actions.

Thus, this chapter examines two fundamental questions:

- Can traditional and social media tools be used to gauge public perceptions about a species that has suffered persecution and needs conservation protection?
- What kind of impacts do mass media and social media have on the conservation of species?

Methods

To answer these questions and understand the interplay among society, science, and policy to conserve species, the authors evaluated four data sources related to hellbender salamanders: peer-reviewed scientific research, online searches, content analysis of newspaper articles, and social media.

To evaluate the public perspectives of hellbenders, the authors analyzed content from print and online news using the LexisNexis Academic database and Newspaper.com for historical newspaper articles, and the Library of Congress Chronicling America database. This content analysis framework was used to evaluate print and online media and to determine whether articles were relatively positive, negative, or neutral in nature (frame). The study then compared article frames across Shaw's Management Eras (Shaw, 1985). It also used Google Trends, following Nghiem, Papworth, Lim, and Carrasco (2016), to assess overall trends from 2004 to 2016 on a scale of 0–100 for interest in Google keyword searches for the term "hellbender." Google Trends provides an avenue to gauge public interest in the species using a common search tool and can be used as a metric for relevant public interest.

To determine how primary scientific peer-reviewed research articles may have affected the authors' observations as another potential source of information on the conservation biology of the species, they enumerated the number of publications annually with either the term "hellbender" in the title or the research subject using Google Scholar (search conducted October 25, 2020). This was performed to assess whether increased research interest (research articles per year as a proxy) was a potential source of information for media coverage. This coverage includes research summaries and quotes from researchers regarding conservation, which is often incorporated into more recent media articles. The Google Scholar search also provided a determination of how the amount of research on this enigmatic species of conservation concern changes over time.

To further assess recent social media and overall social media engagement, the authors conducted a recent search of attitudes towards hellbenders from several online social media outlets via BuzzSumo, similar to Waszak et al. (2018) and Alsyouf et al. (2019). BuzzSumo generates article links for keywords across social media platforms, including Facebook, Twitter, Pinterest, and Reddit. The study defined a social media publication as any material (articles, blogs, memes, videos, etc.) posted on social media. Engagements are represented by how many times people liked, viewed, or shared the posted material. The BuzzSumo search found many unrelated posts that the authors did not count because they were not specifically about the hellbender salamander, including a heavy metal band, a Western movie, cycling equipment, anime, a brewery, fishing lures, an unrelated book, a motorcycle shop, trails, and a highway.

Social media includes forms of electronic communication such as social-networking and blogging sites where individuals can form groups and communicate and share content. Content includes primary, secondary, or self-published articles, news, videos, memes, and additional personal messages. Most social media are thought to have gained a following in the 2000s, including Facebook and Twitter, formed in 2004, and 2006, respectively. Emerging social media provide a valuable tool to promote fact-based natural history, dispelling myths and misconceptions, as well as a method to engage and involve the public in citizen science (Bik & Goldstein, 2013; Toivonen et al., 2019; Bickford, Posa, Qie, Campos-Arceiz, & Kudavidange, 2012; Lo, 2016).

Findings

Overall, this sample of public attitudes until 2016 were positive following the Conservation Biology Era (since 1980) and were reflected in both print and online media. Moreover, social media have seen a large increase in coverage of hellbenders, with an overall increase in engagement, as well as more scientific research about this enigmatic animal. Traditional media were driven by conservation science, which could be associated with social changes (Unger & Hickman, 2020). While traditional media were once the gatekeepers of publishing information for general consumption, information about species conservation can now go directly from scientists to the public via social media. Therefore, scientists should engage in this newer social media format for online media to inform the public.

Historical newspaper content across conservation eras

To determine whether public attitudes towards a species would follow management eras (increasing positive perceptions aligning with conservation biology), this study examined 153 years of media coverage across the hellbender salamander's geographic distribution in the United States. To address our main hypothesis that perceptions about hellbenders will change according to conservation eras over time, we assigned articles a valence value for frame (positive, neutral, or negative) using several criteria (Houston, Bruskotter, & Fan, 2010; Rust, 2015). For example, articles were assigned a positive frame if they mentioned conservation efforts by universities, zoos, or federal or state entities. Articles were scored as positive if they described that hellbenders were important to the environment or ecosystem (ecosystem service), or legislative change for protection status. We assigned articles a neutral frame if they discussed multiple perceptions such that not just one dominated the article. For example, the phrase "hellbenders are ugly, but harmless" sends mixed information about the animal and might not help a reader decide if protection or persecution of the species is needed. Finally, negative frames were assigned to articles that might describe how hellbenders are harmful in some way (e.g., consume fish or may bite you), or with a negative description such as "lazy," "gross," "repulsive," "vile," or "hideous." Negative frames might simply describe hellbenders undesirably (e.g., "beasts" or "ugly") but without a contrasting benefit to the ecosystem or uniqueness that describes them with an endearing quality. A negative article might also describe the need for people to kill hellbenders. In summary, we scored the article frame as positive (1), negative (−1), or neutral (0), and then averaged scores from the articles for each year with the frame as the response variable (Unger & Hickman, 2020). Following a review and categorical placement of each article, the analysis determined whether overall published attitudes towards hellbender salamanders differed across five of Shaw's management eras (Shaw, 1985). The study determined that a positive public sentiment represented by perceptions in newspaper media aligned well with the Conservation Biology Era (Figure 11.3). One example includes changes in how newspapers portrayed hellbenders from 1906. In this example that occurred in Harrisburg, Pennsylvania, an individual hellbender was captured and subsequently "skinned" and eaten by zoology students and prepared as a "culinary experiment." Moreover, the head scientists, according to this article, proclaimed, "There's no reason why the hellbender shouldn't be good to eat." Yet this article further described the hellbender as "a mighty fine fish" and implied its edibility if one could "conquer his natural aversion" – presumably – to its unique morphology and overall appearance. However, the researchers noted a shift in newspaper content over time and from its portrayal as "ugly" to actively promoting and reporting research efforts and evidence of the importance of protecting its habitat and overall conservation. In more recent years, many articles mentioned the value of hellbender salamanders as indicators of water quality – such as being an aquatic version of the "canary in the coal mine" – referring to their ability to indicate threats to human life if they sicken or die.

Google trends

There was a peak interest in "hellbenders" in late 2011, which coincided with the listing of the Ozark hellbender as an endangered species, as well as the

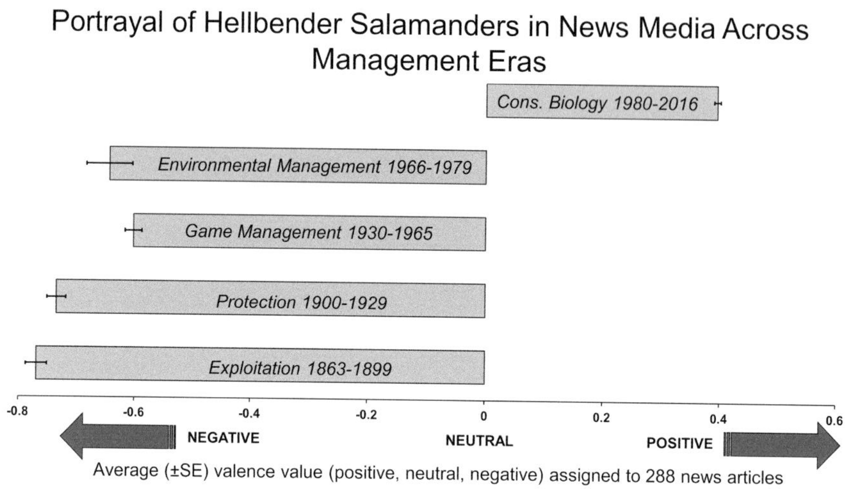

Figure 11.3 Mean valence scores (+/−) for article frame compared across five Conservation Eras. Closer to zero reflects a more neutral value, while positive and negative are closer to 1 and −1 for perception. In other words, if on the positive side, articles generally described hellbender salamanders as helpful in some way. If on the negative side, hellbenders were generally described as destructive or damaging in some way. Neutral scores showed ambivalence or simple curiosity.

Source: Graph recreated from Unger & Hickman, 2020 and published with permission.

successful captive breeding of Ozark hellbenders at the Saint Louis (Missouri) Zoo (Ettling et al., 2013). This peak in interest through online searches can be interpreted as an indication of the media's potential to generate additional interest among the general public after reading media reports on either of these events.

Moreover, the number of scientific research articles on hellbenders found through Google Scholar searches shows over a 250% increase in hellbender research after 2005 (Figure 11.4). However, uncertainty remains as to how this increase in scientific studies translates to increased engagement in social media or online platforms, but the authors suspect it has increased awareness as some researchers were interviewed in newspapers and online media formats that were subjects of popular social media posts.

Social media engagements and publications

This study monitored one year (from October 18, 2019, to October 17, 2020) of social media engagements and publications across four major social media platforms: Facebook, Twitter, Pinterest, and Reddit. Almost all results centered on hellbender research, discovery, and conservation. The search identified 92 publications, from a variety of outlets on social media, including news sources,

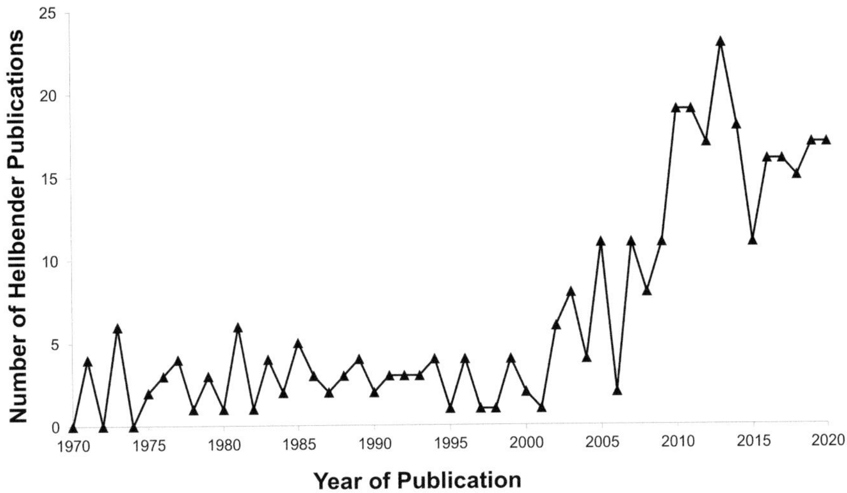

Figure 11.4 Annual number of publications with the terms "hellbender" or "hellbender salamanders" in the title or the primary emphasis of research articles on either Eastern or Ozark hellbender species.

Source: Graph courtesy of Caleb R. Hickman and Shem D. Unger.

blogs, videos, memes, and even primary research articles that were sharable and had a searchable title or subject line for "hellbender" or "hellbender salamanders." Over social media, there were 90,143 engagements, which is the sum of all likes, comments, and shares about individual publications (also considered posts). Therefore, through social media, single publications can find new life and can reach a variety of outlets, and thus a greater proportion of the public. For example, a traditional daily newspaper, the Independence, Missouri *Examiner*, has print and online formats. On October 5, 2020, the *Examiner* published an article titled "Rare hellbenders released in the Ozarks," which had 36 Facebook engagements (shares, comments, likes, and posts) and three Twitter shares. Other examples of publications come from videos, such as one created for YouTube titled "In search of the Eastern hellbender salamander," which resulted in 51 Facebook engagements and three Twitter shares.

From this monitoring effort, the authors reviewed and enumerated online material, such as news articles, videos, and memes posted on social media and how many people accessed this material by viewing and sharing it, which are considered separate engagements. The authors defined posts as separate instances where individual people shared information about hellbenders in the form of articles, videos, and memes, and defined engagements as each time an individual read material on a particular social media format. The most impactful month for social media engagement was October 2019, with 43,232 engagements over six posts (Figure 11.5). However, during the sample year, 19

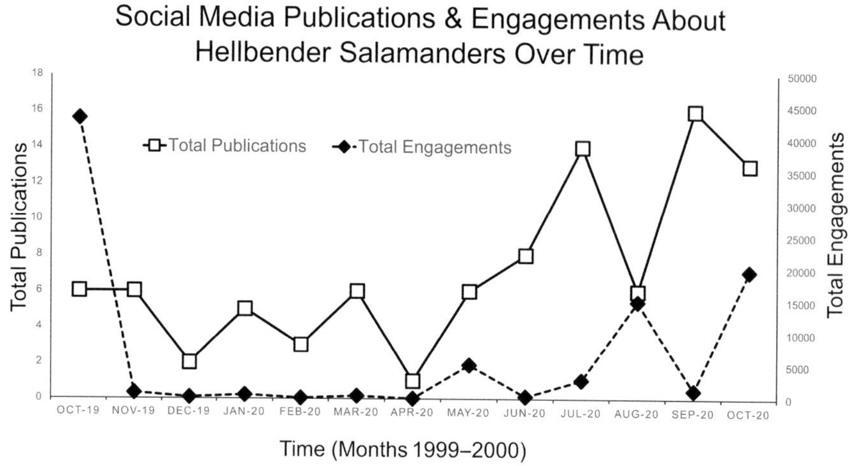

Figure 11.5 Total social media engagements and social media publications for the terms "hellbender" and "hellbender salamanders" for October 2019–October 2020. Publications (open square) are represented by the number of individual media, such as news articles, primary research articles, videos, magazines, blogs, or memes in a month. Engagements (dark diamond) are represented by the number of ways that people can interact with a single publication (post), including the sum of likes, shares, and comments for each month.

Source: Graph courtesy of Caleb R. Hickman and Shem D. Unger.

posts received no engagements and 150 posts had 90,144 engagements across all outlets (Figure 11.5).

The top three posts in the number of engagements were: "Ohio resource officers release hellbenders in Ohio waters" (newspaper article with 13,600 engagements), "Hellbenders released into Ozark rivers" (television news story and article with 19,600 engagements), and "Hellbenders found in a new Pennsylvania water" (newspaper article with 23,500 engagements). Most engagements came from the website YouTube.com (3,805), followed by Anchor.fm (78), and Goodreads (6). By far, Facebook had the most engagements (89,151 over 64 publications), while Twitter hosted more publications (550 shares over 81 separate publications). Since Facebook allows users to react to publications (posts) through emojis, the platform provided a look at the relative percentage of different reactions based on emojis that users selected for each publication. To display the results, the authors rounded reactions to the nearest whole percentage. Reaction emojis included "Angry," "Haha," "Love," "Sad," and "Wow." These reactions could be responses to a variety of stories about the conservation work, discovery, unusual morphological appearance of hellbenders, or to the sad state of the hellbenders that are at risk of endangerment. Almost all reactions were positive (98%), as represented by emojis of "Love," "Haha," and

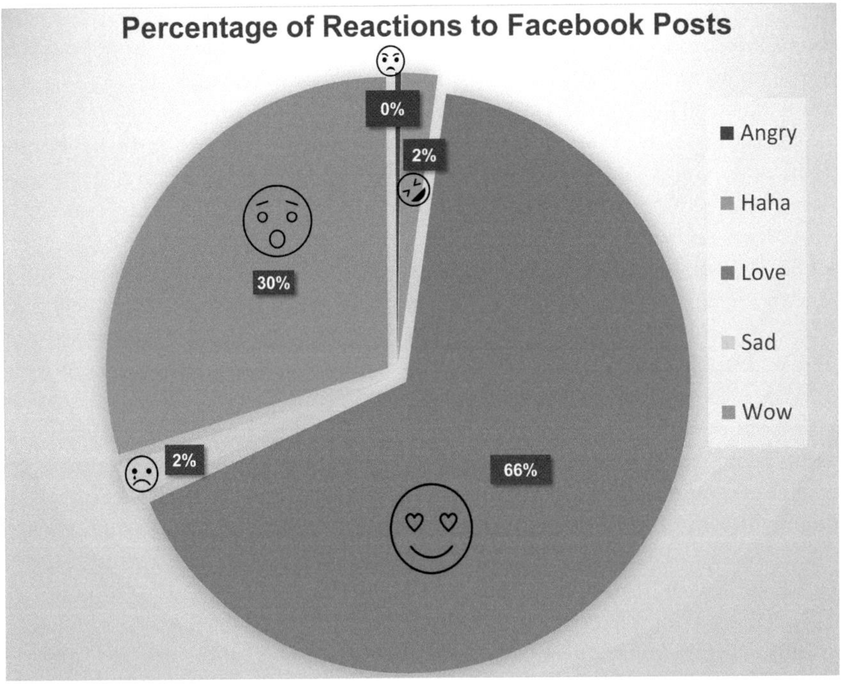

Figure 11.6 Public reactions displayed through emojis to Facebook posts for October 2019–2020. More negative meanings: "Angry," pouting face shows upset intentions towards the publication; "Sad," crying face with animated tear means the publication initiated a feeling of sadness. More positive meanings: "Haha," laughing face means the publication is humorous; "Love," beating heart means the publication initiates a feeling of caring; "Wow," surprised face shows reader received new information with excitement.

Source: Graph courtesy of Caleb R. Hickman and Shem D. Unger.

"Wow" (Figure 11.6), leading the authors to believe that most posts relating to conservation and discovery were well-received by the public.

Discussion and conclusions

Hellbenders are a species that, historically, engendered negative perceptions but have gained more public interest and appeal in recent years. The USFWS denied the Eastern hellbender federal endangered species status in 2019, with the agency stating that some populations were apparently healthy within their geographic range. Researchers met this reasoning – and the resulting decision – with mixed reactions. Moreover, a petition is currently pending with the USFWS to list the Missouri population as distinct, allowing it to be listed as a separate species within the state and granting Missouri hellbenders their own federal protection. Could social support influence federal designation? Could the public be a voice for more legislation to promote research money from the US or state governments (via the Wildlife Action Plan or newer Recovering America's Wildlife Act)?

While this study noted an increase in the positive perception of hellbenders in newspapers, along with an associated increase in social media postings over time, further work needs to be done across states to promote conservation efforts and inform the public. From this work, conservationists can gain public "buy-in" to strengthen support for science-based management of hellbenders, as much conservation efforts for this species occur at the local and state levels and would benefit from public support. The authors' observation of increased publications on hellbenders over time correlates with media coverage they observed in the Conservation Era. The authors observed that some traditional newspapers, as well as online social media, included descriptions of researchers, discovery, and conservation. These platforms often communicated local conservation efforts by researchers, as well as environmental threats to the species. The diverse forms of social media technologies and the ease of their use can encourage scientists to increase communication of science through more traditional forms of dissemination, with media professionals being historically the gatekeepers of scientific information (Dudo, 2015). Social media will likely continue to provide a valuable tool to inform the public about specific conservation issues.

While social media can provide a forum for increasing public awareness of wildlife conservation (Wu et al., 2018), there remains work to be done in this new digital age where social media can influence the public regarding conservation. Figure 11.7 illustrates how scientists' use of social media can affect public perception. Researchers should use a variety of outlets to inform the public on conservation of hellbenders and other species of conservation concern. While scholarly journals have begun to promote their research articles to social media platforms (Zheng et al., 2019), further engagement by individual conservation biologists at both local and regional levels may potentially impact community landowners and the public at large. Moving forward, a greater emphasis on communicating accurate information on conservation efforts and species natural history, along with population trends, should encourage public buy-in while helping to dispel myths and misconceptions. However, misinformation about science can be magnified on social media (Baker, 2019). It is, therefore, incumbent on conservation researchers to acquire greater skills in communicating via social media and to become more proactive to increase environmental literacy among both the public and policy makers (Bickford et al., 2012).

It is clear that the Conservation Era (1980–present) coincided well with positive perspectives about hellbenders. This could be for several reasons. The Conservation Era followed a large policy change in favor of funding and protecting a variety of non-game species. It is likely that scientists' conservation

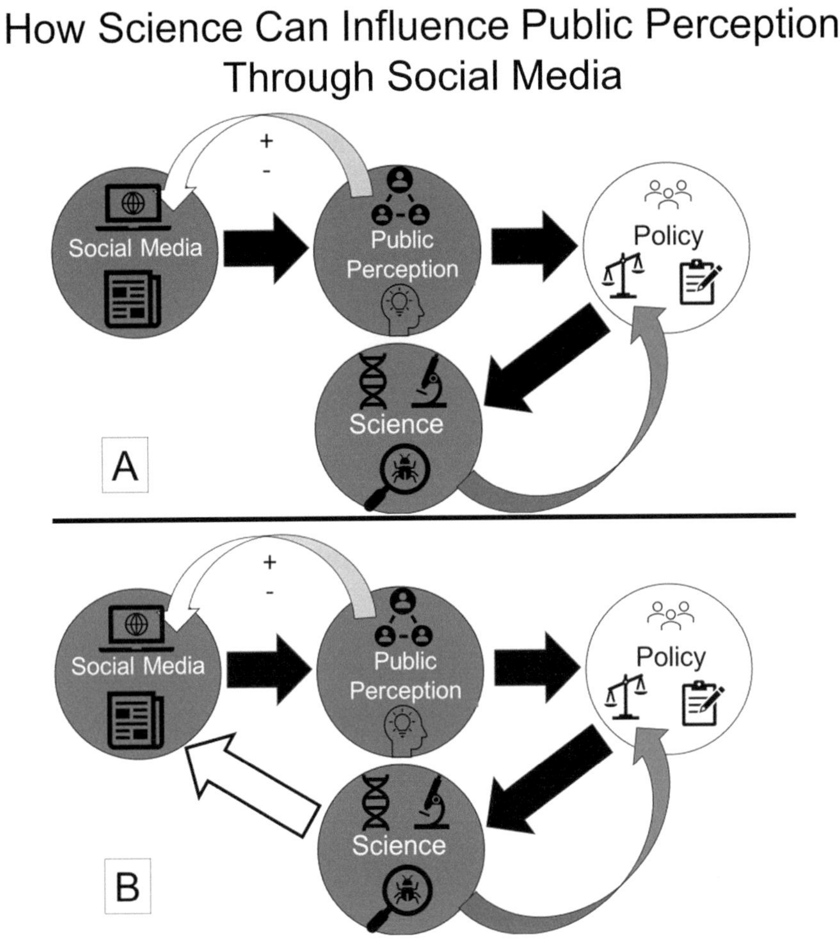

Figure 11.7 Comparative conceptual diagrams show how scientists can influence public perception if using social media. Darker circles are where social information is developed. With a shift to a fast-paced online social media format to drive information, the traditional science-to-media process can be replaced with public perception instead of that from science (A). If public perception is the only information source, it can develop both positive and negative perspectives in a feedback loop that is unsupported by science (clear arrow – A). In the lower diagram (B), science information is added (light arrow) to social media, which can be shared across platforms, thereby influencing decisions. After sharing on social media, information is processed through the public filter making requests to policy makers (light circle).

Source: Image courtesy of Caleb R. Hickman and Shem D. Unger.

research on hellbenders generated more information to be publicized, as indicated by the number of research articles published and available on Google Scholar (Figure 11.4) and the number of social media publications and user engagements (Figure 11.5). The Google Trends results (Figure 11.3) showed a large increase in searches following the well-documented hellbender breeding and restoration work by the Saint Louis Zoo, which followed the ESA listing of the Ozark subspecies in 2011. As of late 2020, the St. Louis Zoo had released more than 1,000 Ozark hellbenders into Ozark streams, which was recognized as one of the most impactful social media events with eight news articles and 20,415 engagements. These conservation actions have the potential to lead to changes of public perception if disseminated by media outlets and shared across social media.

The most important questions for future exploration are whether public perceptions help drive policy change and if social media can initially drive such public perceptions. With more information than ever generated by scientists and secondary sources sharing across social media (Toivonen et al., 2019), it could be said that we are entering the Social Media Conservation Era. For example, while listing and delisting species such as the gray wolf (recently delisted by the USFWS as an endangered species) is not without political and policy disputes (Hamilton et al., 2020), social media stand as a powerful vehicle to promote conservation and environmental communication. There is evidence to support scientists taking a more active role in informing the public through both traditional and social media (Bik & Goldstein, 2013; Wu et al., 2018; Toivonen et al., 2019; Veletsianos & Shaw, 2018). Moreover, it is vital for scientists to communicate their research and scientific information to the public via social media (Eperen & Marincola, 2011), as it likely will affect public perception of actual science and, in the long run, accurate portrayal of conservation. Without input from scientists, information on social media might be poorly informed by common misconceptions and biased perceptions of species of conservation concern (Figure 11.7A). If the media perpetuate misconceptions about a species, the public may believe these misconceptions, leading to misinformation spreading among the public. This public misinformation has the potential to misinform policy decisions. However, if scientists are more involved and disseminate accurate primary information and actively engage on social media platforms, common misconceptions can be minimized or removed from the policy making process (Figure 11.7B). Researchers should create online summaries of their work and help the public understand the process of science through online platforms and reviewing the work of their colleagues. Conservation biologists must embrace social media as a form of outreach education in an age where online information has the potential to influence a large segment of the population. Based on the number of engagements from publications (posts), this study concluded that scientists should place conservation material on social media platforms like Facebook and YouTube. A practice of engaging the public through popular media gives scientists an opportunity to dispel old stereotypes of esoteric science and influence policy through public participation. When engaging on social media or in comment areas of a publication, scientists can be important fact-checkers for work in their field. Furthermore, researchers can use social media analysis, as was done here, to evaluate whether their science is generating any influence in society and to assess how they might promote changes in perceptions, and thereby potentially influence policy and conservation.

Acknowledgments

This chapter draws on research previously published in Unger, S.D., & Hickman, C.R. (2020). A content analysis from 153 years of print and online media shows positive perceptions of the hellbender salamander follow the conservation biology. *Biological Conservation, 246*, doi:10.1016/j.biocon.2020.108564. The authors thank the Eastern Band of Cherokee Indians and Wingate University for resources and salary to pursue this research.

References

Alsyouf, M., Stokes, P., Hur, D., Amasyali, A., Ruckle, H., & Hu, B. (2019). "Fake news" in urology: Evaluating the accuracy of articles shared on social media in genitourinary malignancies. *BJU International, 4*, 701–706.

Baker, B. (2019). Biologist as communicators: Lessons from social science can help researchers get their message across. *BioScience, 69*, 101–107.

Bickford, D., Posa, M. R. C., Qie, L., Campos-Arceiz, A., & Kudavidange, E. P. (2012). Science communication for biodiversity conservation. *Biological Conservation, 151*, 74–76.

Bik, H. M., & Goldstein, M. C. (2013). An introduction to social media for scientists. *PLoS Biology, 11*(4), e1001535. doi:10.1371/journal.pbio.1001535

Brossard, D., & Scheufele, D. A. (2013). Science, new media, and the public. *Science, 339*, 40–41.

Burgmeier, N. G., Unger, S. D., Sutton, T. M., & Williams, R. N. (2011). Population status of Eastern hellbender (*Cryptobranchus alleganiensis alleganiensis*) in Indiana. *Journal of Herpetology, 45*, 195–201.

Caniglia, R., Galaverni, M., Delogu, M., Fabbri, E., Musto, C., & Randi, E. (2016). Big bad wolf or man's best friend? Unmasking a false wolf aggression on humans. *Forensic Science International: Genetics, 24*, e6.

Costas, R., Honk, J., & Franssen, T. (2017, October 16–20). Scholars on Twitter: Who and how many are they? 16th International Conference on Scientometrics and Informetrics, Wuhan, China.

Darwin, C. (1859). *On the origin of species by means of natural selection, or, the preservation of favoured races in the struggle for life*. London: J. Murray.

Dary, D. A. (1989). *The buffalo book*. Athens, OH: Swallow Press, Ohio State University Press.

Dudo, A. (2015). Scientists, the media, and the public communication of science. *Sociology Compass, 9*(9), 761–775.

Dunaway, W. (1994). The Southern fur trade and the incorporation of Southern Appalachia into the world-economy, 1690–1763. *Review (Fernand Braudel Center), 17*(2), 215–242. Retrieved from www.jstor.org/stable/10.2307/40241289

Eeden, L., Dickman, C., Ritchie, E., & Newsome, T. (2017). Shifting public values and what they mean for increasing democracy in wildlife management decisions. *Biodiversity and Conservation, 26*, 2759–2763.

Eperen, L. V., & Marincola, F. M. (2011). How scientists use social media to communicate their research. *Journal of Translational Medicine, 9*, 199.

Ettling, J. A., Wanner, M. D., Schuette, C. D., Armstrong, S. L., Pedigo, A. S., & Briggler, J. T. (2013). Captive reproduction and husbandry of adult Ozark hellbenders. *Cryptobranchus alleganiensis bishopi: Herpetological Review, 44*, 605–610.

Fogelman, V. M.(1988). American attitudes towards wolves: A history of misperception. *Environmental Ethics, 10*, 63–94.

Hamilton, L. C., Lambert, J. E., Lawhon, L. A., Salerno, J., & Hartter, J. (2020). Wolves are back: Sociopolitical identity and opinions on management of *Canis lupus. Conservation Science and Practice*, 1–13. doi:10.1111/csp2.213

Hillis, R. R., & Bellis, E. D. (1971). Some aspects of the ecology of the hellbender, Cryptobranchus alleganiensis, in a Pennsylvania stream. *Journal of Herpetology, 5*, 121–126.

Houston, M. J., Bruskotter, J. T., & Fan, D. (2010). Attitudes toward wolves in the United States and Canada: A content analysis of the print and news media. 1999–2008. *Human Dimensions of Wildlife, 15*(5), 389–403. doi:10.1080/10871209.2010.507563

Howell, E. L., Nepper, J., Brossard, D., Xenos, M. A., & Scheufele, D. A. (2019). Engagement present and future: Graduate student and faculty perceptions of social media and the role of public in science engagement. *PLoS One, 14*(5), e0216274. doi:10.1371/journal.pone.0216274

Kallman, H. (Ed.). (1987). *Restoring America's wildlife, 1937–1987: The first 50 years of the Federal Aid in Wildlife Restoration (Pittman-Robertson) Act*. Falls Church, VA: US Department of the Interior, Fish and Wildlife Service.

Lo, Y. (2016). *Online communication beyond the scientific community: Scientists use of new media in Germany, Taiwan, and the United States to address the public* (PhD dissertation). Freie Universitat, Berlin.

Mullendore, N., Mase, A. S., Mulvaney, K., Perry-Hill, R., Reimer, A., Behbehani, L., . . . Prokopy, L. S. (2014). Conserving the Eastern hellbender. *Human Dimensions of Wildlife, 19*, 166–178.

Musiani, M., & Paquet, P. C. (2004). The practices of wolf persecution, protection, and restoration in Canada and the United States. *BioScience, 54*(1), 50–60.

Netting, M. G. (1929). The food of the hellbender, Cryptobranchus alleganiensis (Daudin). *Copeia*, 23–24.

Nghiem, L. T. P., Papworth, S. K., Lim, F. K. S., & Carrasco, L. R. (2016). Analysis of the capacity of Google trends to measure interest in conservation topics and the role of online news. *PLoS One, 11*(3), e0152802.

Nickerson, M. A., & Briggler, J. T. (2007). Harvesting as a factor in population decline of a long-lived salamander, the Ozark hellbender, *Cryptobranchus alleganiensis bishopi* Grobman. *Applied Herpetology, 4*, 207–216.

Nickerson, M. A., & Mays, C. E. (1973). A study of the Ozark hellbender *Cryptobranchus alleganiensis bishopi. Ecology, 54*(5), 1164–1165.

Pandey, D. P., Pandey, G. S., Devkota, K., & Goode, M. (2016). Public perceptions of snakes and snakebite management: Implications for conservation and human health in Southern Nepal. *Journal of Ethnobiology and Ethnomedicine, 12*, 22.

Perry-Hill, R., Smith, J. W., Reimer, A., Mase, A. S., Mullendore, N., Mulvaney, K. K., & Prokopy, L. S. (2014). The influence of basic beliefs and object-specific attitudes on behavioral intentions towards a rare and little-known amphibian. *Wildlife Research*, *41*, 287–299.

Petranka, J. W. (1998). *Salamanders of the United States and Canada*. Washington, DC: Smithsonian Institution Press.

Phillips, C. A., & Humphries, W. J. (2005). *Cryptobranchus alleganiensis* (Daudin, 1803). Hellbender. In M. Lanoo (Ed.), *Amphibian declines: The conservation status of United States species* (pp. 648–651). Berkeley, CA: University of California Press.

Portman, M. E. (2013). Ecosystem services in practice: Challenges to real world implementation of ecosystem services across multiple landscapes – a critical review. *Applied Geography*, *45*, 185–192. doi:10.1016/j.apgeog.2013.09.011

Purcell, K., Brenner, J., & Rainie, L. (2012). *Search engine use 2012*. Washington, DC: Pew Internet & American Life Project.

Reimer, A., Mase, A., Mulvaney, K., Mullendore, N., Perry-Hill, R., & Prokopy, L. (2014). The impact of information and familiarity on public attitudes towards the Eastern hellbender. *Animal Conservation*, *17*, 235–243.

Roberge, J. (2014). Using data from online social networks in conservation science: Which species engage people the most on Twitter? *Biodiversity and Conservation*, *23*, 715–726.

Rossell, C. R., McNeal, P., Gillette, D. P., Williams, L. A., Patch, S. C., & Krebs, A. G. (2013). Attributes of shelters selected by Eastern hellbenders (Cryptobranchus a. alleganiensis) in the French Broad River Basin of North Carolina. *Journal of Herpetology*, *47*(1), 66–70.

Rust, N. A. (2015). Media framing of financial mechanisms for resolving human-predator conflict in Namibia. *Human Dimensions of Wildlife*, *20*(5), 440–453.

Shaw, J. H. (1985). *Introduction to wildlife management*. New York: McGraw-Hill.

Smith, K. F., Behrens, M., Schloegel, L. M., Marano, N., Burgiel, S., & Daszak, P. (2009). Reducing the risks of the wildlife trade. *Science*, *324*(5927), 594–595. doi:10.1126/science.1174460

Soule, M. E. (Ed.). (1985). What is conservation biology? *Bioscience*, *3*, 727–734.

Toivonen, T., Heikinheimo, V., Fink, C., Hausmann, A., Hiippala, T., Jarv, O., . . . Di Minin, D. (2019). Social media data for conservation science: A methodological overview. *Biological Conservation*, *233*, 298–315.

Unger, S., & Hickman, C. (2020). A content analysis from 153 years of print and online media shows positive perceptions of the hellbender salamander follow the conservation biology. *Biological Conservation*, *246*, 108564.

Veletsianos, G., & Shaw, A. (2018). Scholars in an increasingly open and digital world: Imagined audiences and their impact on scholars' online participation. *Learning, Media and Technology*, *43*(1), 17–30. doi:10.1080/17439884.2017.1305966

Waszak, P. M., Kasprzycka-Waszak, W., & Kubanek, A. (2018). The spread of medical fake news in social media – the pilot quantitative study. *Health Policy and Technology*, *7*, 115–118.

Wheeler, B. A., Prosen, E., Mathis, A., & Wilkinson, R. F. (2003). Population declines of a long-lived salamander: A 20+- year study of hellbenders, *Cryptobranchus alleganiensis*. *Biological Conservation*, *109*, 151–156.

Wu, Y., Xie, L., Huang, S., Li, P., Yuan, Z., & Liu, W. (2018). Using social media to strengthen public awareness of wildlife conservation. *Ocean and Coastal Management*, *153*, 76–83.

Zheng, H., Aung, H. H., Erdt, M., Peng, T., Raamkumar, A. S., & Theng, Y. (2019). Social media presence of scholarly journals. *Journal of the Association for Information Science and Technology*, *70*, 256–270.

Part IV

Environmental public policy

12 (Not) saving the Great Barrier Reef from disaster

Media then and now

Kerrie Foxwell-Norton and Libby Lester

The Great Barrier Reef is an internationally recognized wonder of the natural world – a global nature superstar. Its scientific credentials are exceptional, with a list of marine creatures that includes 600 types of soft and hard corals, more than 100 species of jellyfish, 3,000 varieties of mollusks, 500 species of worms, 1,625 types of fish, 133 varieties of sharks and rays, and more than 30 species of whales and dolphins. The Great Barrier Reef Marine Park covers 133,000 square miles (344,400 square kilometers) and contains the world's largest coral reef ecosystem with 3,000 reefs, 600 continental islands, 300 coral cays, and 150 inshore mangrove islands. The Reef is part of Australia's national identity, with Australians defining themselves as coastal dwellers "living on the edge" (Drew, 1994), and it is a site of historical and contemporary Indigenous cultural heritage, retaining significance to Australia's First Nations people, especially those whose traditional lands border the Reef. It is the location of a brutal colonial history, invasion, and frontier encounters, also part of the Australian identity (McCalman, 2013). In 2017, the Great Barrier Reef was valued at $56 billion (AU) as an Australian economic, social, and iconic asset. In 2015–2016, the Reef contributed $6.4 billion to the Australian economy and added over 64,000 jobs (direct and indirect) to the national economy (Deloitte Access Economics, 2017). Stretching for 1,550 miles (2,500 kilometers) along the Queensland coast, it supports numerous regional towns, communities, and businesses, particularly those reliant on tourism and fishing, and thus the Reef continues its status as a holiday destination.

Statement of the problem

This chapter examines two distinct periods of conflict in the history of this nature superstar: the Save the Reef campaign of the 1960s and 1970s and the Fight for the Reef campaign that began in 2012. These were both dedicated to conservation of the Reef and, principally, protection from the nefarious impacts of fossil-fuel extraction – oil and gas exploration in the earlier campaign and thermal coal in the latter. Media communications played an important role in both, and this study locates and analyzes specific features of these campaigns within the context of major change to media practices, logics, and technologies – alongside dynamic social and political settings. In doing so, this chapter contributes to important debates within media and communications – that is, the bounds and opportunities for the public to participate in debates on shared environmental futures, and the potential impact of such campaigns and debates on the formation of policy geared to environmental protection and conservation. Building on previous work (Foxwell-Norton & Lester, 2017), and in the context of ongoing threats to Reef health, especially from climate change and rising sea temperatures, the study evaluates the contemporary media landscape's potential to save – or not – the Great Barrier Reef.

Both campaigns focused on preventing the destruction of the Reef. In 1967, Australian artist and environmental campaigner John Büsst read a public notice in a small community newspaper that a local farmer had applied to mine limestone on Ellison Reef, offshore and about midway between the small Queensland communities of Innisfail and Mission Beach. Local sugar cane growers wanted to use limestone from the Ellison Reef coral as fertilizer, which would have required little transportation or further treatment. Subsequently, local news media revealed and reported proposals for oil and gas exploration, specifically noting the pecuniary interests of senior politicians (Wright, 1977). After seven years of protest and political maneuvering (see Hutton & Connors, 1999), the Great Barrier Reef became a marine park in 1974 with a World Heritage listing following in 1981. The story of the proposal to mine limestone on Ellison Reef and its implications for Reef protection has been told in various forums, most focused on surrounding politics and protest (Clare, 1971; Wright, 1977; McCalman, 2013; Bowen & Bowen, 2002; Evans, 2007; Hutton & Connors, 1999). This chapter offers another dimension by considering the strategic and media communications of the Save the Reef campaign and exploring the campaign's contribution to conservation efforts.

Fight for the Reef launched in 2012 as a partnership between the Australian Marine Conservation Society and World Wildlife Fund-Australia in response to major development proposals. These proposals had also prompted a World Heritage monitoring mission because they included a massive expansion of coal mining and associated activities in Queensland, with nine mines proposed in the Galilee Basin deposit 250 miles (400 kilometers) inland from the Reef. The Carmichael Mine, owned by an India-based company, the Adani Group, planned to produce 60 million tons of coal a year (Adani, 2014). MacMines Austasia, now solely owned by the Meijin energy group, one of China's largest producers of coal, was approved to produce 70 million tons of coal annually, with an expected mine life of 40 years. Coal from both mines, along with others awaiting approval, would be transported to expanded shipping facilities at Abbot Point, on the central Queensland coast, where dredging was proposed to allow ships transporting the coal to make their way through the Reef. Legal action and a change of state government left the status of the proposal unclear, while governments at both state and federal levels continued to reassure investors of their commitment to the Queensland coal industry. In 2019, Adani, now rebadged as Bravus Mining and Resources, received its final environmental approvals and was able to commence work on the Carmichael Mine. Controversy persists surrounding its current and future levels of production and expected employment opportunities (Australian Broadcasting Corporation, 2019) as legal challenges attempted to delay the project and government decision-making (see Konkes, 2018). Pressure brought to bear on financing the Adani mine has been a feature of recent protests, with campaigns targeted at banks and insurance companies underwriting the mine project. Saving the Great Barrier Reef remains subject to international scrutiny, heightened by reporting of its "death" followed by media encouraging "last chance visits" (see McGaurr & Lester, 2018), which affected engagement with efforts to mitigate climate change impacts on the Reef (Eagle, Hay, & Low, 2018). These international networks and flows of local to global communication and media brought the spectacle of Reef protest to audiences near and far – demanding that industry and governments navigate these local and transnational publics (Lester, 2019).

Findings

To begin outlining the political and social contexts and a conceptual framework for this study, it is useful to emphasize that although media landscapes of today seem barely related to those of 1967, media have in fact played a long and continuing role in the communication of environmental risks and concerns. Media centrality in communication of a range of environmental issues, events, and places and the negotiation of shared environmental futures is well recognized (Hansen, 2010; Cox, 2012; Lester, 2010). This case study reinforces that centrality. Since the Reef's "discovery" by European explorers and early Australian colonial settlers, media have played a critical role in communicating knowledge about the Reef to the world. Today, despite the relatively high number of visitors to the region, few people have directly experienced the Reef, in particular the outer Reef from where our visual representations largely emanate. The fact that a notice in Büsst's local newspaper alerted him to the threat to mine the Great Barrier Reef is also salient.

However, at the time Büsst was reading of the plans to mine Ellison Reef, crucial transformations were underway. In 1960s regional Australia, email and the internet were decades away, and translocal communication was mostly by an often-inefficient mail system and expensive telephone calls. Television was barely a decade old in metropolitan areas and still unavailable in large parts of regional Australia. Color television would not be introduced to Australian homes until the mid-1970s. Meanwhile, images of the Earth from space from the Apollo 8 mission in 1968 were greeted as remarkable – a "communications moment" that some scholars have identified as signaling the birth of the modern environmental movement (Cosgrove, 1984; Lester, 2010). The mission also involved a radio broadcast and, in the words of Apollo astronaut Frank Borman: "We were told that on Christmas Eve we would have the largest audience that had ever listened to a human voice" (NASA, 2014). In this sense, a simultaneous moment of global environmental communication and consciousness occurred, although such a moment was possible only because of a collective awareness and movement towards care for the environment and associated politics. This included, for example, the publication of Rachel Carson's *Silent Spring* in 1962 (which reached Australian shores as a campaign tool for activists wanting to highlight the threat of industrial production and processes), designation of dedicated environment reporters by Australian newspapers, and the loss of picturesque Lake Pedder in Tasmania's remote southwest after a protracted nationwide battle against a hydroelectric development (Lester, 2010, 2014).

The first environmental policies designed to address the consequences of large-scale industrial production and development began to appear. In the United States, environmental policies to police industrial production, particularly nefarious impacts on water and air quality, emerged. On an international scale, in 1968 Sweden suggested that the UN Economic and Social Council hold a conference focused on human interactions and the environment. The proposal eventually led to the UN Stockholm Conference in 1972 and its focus on environment and development. International policy also gained significant momentum with the establishment of a program to designate World Heritage Sites at the 1972 Paris UNESCO Conference, thus delivering a mechanism to protect global natural and cultural places of outstanding value. In 1974, Australia was one of the first nations to ratify the World Heritage Convention before it came into force in 1975 (Australian Government, n.d.).

A 1981 meeting of the World Heritage Committee in Sydney, chaired by then-Prime Minister Malcolm Fraser, inscribed Australia's first World Heritage properties: Kakadu National Park, the Willandra Lakes Region, and the Great Barrier Reef. The United Nations Convention on the Law of the Sea, while not passed until 1982, also gained momentum during the 1960s with enactment of four treaties that further defined state sovereignty and territorial waters. Environmental policy and law became the formal mechanisms at local, national, and international levels to administer and safeguard environments.

Alongside this growing global environmental consciousness, communications facilitated global economic growth. While this matrix of law and policy to protect environments in the context of global communications and money emerged during the earlier Save the Reef campaign, it was fully instituted 50 years later during the Fight for the Reef campaign.

Increased awareness of risks posed by industrialization and pressure to mitigate environmental harms have been identified as defining features of the later period of this study (Beck, 2009). Escalating evidence and accompanying concern about the threat of industrial practices on environments – accessed via new forms and flows of information and images – have transformed the ways in which the world (or at least those with access to communication technologies) understands environments. The complexity of the terrain fits within Beck's (2009) concept of "relations of definitions" – the range of opinions and claims that frames the way an issue or risk travels into the public domain for debate. It is, according to Cottle, a "matrix of interests and ideas and the different rationality claims (both scientific and social) that publicly define, contend, and challenge within the [environmental] risk field including the media" (Cottle, 2006, p. 123). Recent theorizing of media and environmental conflict in a networked society has argued that media are both "performative" and "constitutive." That is, the media provide the actions and arenas whereby issues are communicated, and they are key actors in defining those issues (Lester & Cottle, 2009). Environmental conflicts are an important site for the recognition of these processes (Hutchins & Lester, 2015). As such, communications and the media have become "inextricably infused in and are indispensable to protests, publics and media participation" (Lester & Cottle, 2015, p. 108).

Cottle (2014) identified six characteristics of media and communications, which he argues are "historically unparalleled" and point to the "extensity and intensity" of the contemporary communications environment and, in particular, its potential to mediatize disasters – disasters that at the current historical juncture are indicative of global crises no longer contained within nation states. These are:

- Scale: information can be transmitted simultaneously on a global scale;
- Speed: time has collapsed, undermining traditional methods of information management;
- Saturation: universalizing means of communication (e.g., mobile phones) with accompanying expectations of access and preparedness to use;
- Social relations enfranchisement: or universalizing technologies that communicatively expand and enfranchise social relations;
- Surveillance: keeping disasters "out of sight" or off the political radar is unlikely due especially to satellite monitoring of governments and civil society actors;
- Visibility: we can not only read and hear, but also can see disaster and global crises.

Campaigns surrounding the Great Barrier Reef illustrate these observations. The Reef is – as the World Heritage Committee, Great Barrier Reef Marine Authority, and environmental campaigns agree – a pinnacle example of an environment visibly on the edge of disaster as a consequence of factors not contained within state borders, primarily climate change but also international markets and industry pressure for continuing development that impacts Reef health. Yet, little is known about the capacity to raise public consciousness and, therefore, to incite reparatory actions within these conditions. A sobering observation is that the potential for change and challenge remains open principally to those in established positions of power (Lester, 2019; Lester & Hutchins, 2009, 2013). Questions must continue to be framed that interrogate apparent and enduring power structures and discourses typical of industrial society, alongside the potential of new media ecologies to foster and develop positive environmental futures and sustainable outcomes for the Reef.

To advance the framing of these questions, this chapter revisits accounts of the Save The Reef campaign (especially Wright, 1977) alongside a contemporary investigation of the Fight for the Reef campaign, focusing on the role of media and communications in raising public awareness of the Reef in danger. Comparing protest actions half a century apart provides insight into the changing relationship between environmental protest and communication media and the potential for protest and media to incite public participation in Reef decision-making and policy direction towards conservation and protection. The next sections detail key campaign moments and contexts to examine the role and efficacy of media and communications in challenging industrial threats to the Reef.

Save the Reef

The phrase "Save the Reef" was initially adopted in the 1960s by a group that formed the Save the Reef Committee, which, according to Wright (1977, p. 84), helped existing conservationist groups such as the Queensland Wildlife Preservation Society. This study uses the phrase to describe the overall network of individuals, organizations, and events that emerged in the late 1960s to campaign on behalf of the Reef. The phrase is also emblematic of the overall campaign period when Save the Reef bumper stickers became widespread. With its genesis in local community action, it eventually became a national movement with a number of newly established environmental organizations and a collection of individuals mobilizing to protect the Reef. Threats included mining, destructive infestations of crown of thorns starfish (whose presence was believed to be a consequence of the commercial and tourist removal of its chief predator, the triton shell), and oil spills (with fears of disaster driven by the Santa Barbara, California, oil spill in 1969). Two organizations in particular figured prominently in this period: the Queensland Wildlife Preservation Society and the Queensland Littoral Society.

The Preservation Society was founded in 1962 by distinguished poet Judith Wright, publisher Brian Clouston, and others as a community-based, nonprofit organization. John Büsst formed a branch in the town of Innisfail in 1966. Communication and media were initial priorities, with Clouston – as owner of Jacaranda Press – offering to publish a magazine that would collect ideas of society founders and members into a single volume for a larger audience. *Wildlife Magazine*, still in publication, became an initial focus of the society. The Preservation Society's access to good writers was a critical advantage in mounting the campaign to save the Reef. In contrast, the Littoral Society (now known as the Australian Marine Conservation Society and a formal partner in the current Fight for the Reef campaign) was formed in 1965 in Brisbane by scientists and students from the University of Queensland and Australia's premier scientific research body, the Commonwealth Scientific and Industrial Research Organization. Prominent in its ranks was Len Webb, at the time one of the country's few ecologists. The discipline was much-maligned by the more established sciences in the early 1960s, but its rise and recognition throughout the decade with its basic premise of interconnectedness of ecosystems proved critical to the campaign against mining of Ellison Reef and the entire Reef. Some assessments of the Reef had argued that "dead coral" could be mined, but Webb and others successfully challenged that idea from the legitimate space of "scientific knowledge" (see Bowen & Bowen, 2002; McCalman, 2013; Hutton & Connors, 1999; Wright, 1977). Eventually, the unlikely trio of poet Wright, forester/ecologist Webb, and artist Büsst came together to lead the campaign (McCalman, 2013).

Communication of the Reef in crisis in the 1960s occurred across multiple platforms. The Save the Reef campaign was built upon mainstream media platforms – radio, print, and television – supported by alternative media in the form of *Wildlife Magazine*, newsletters, and community media such as local radio and newspapers. Interpersonal and organizational communications among the chief protagonists were also important, although the speed was somewhat limited by the postal services and telephony available at the time. Nevertheless, Büsst dispatched 4,000 letters across the globe during his campaign (Queensland Heritage Register, n.d.).

Mainstream media were critical to the success of Save the Reef with news coverage consistently sought and triumphantly celebrated. Wright understood the power of media, as this excerpt from her 1977 foreword to *The Coral Battleground* illustrates:

> I wish to acknowledge the role of the press, both in reporting and in producing special articles and surveys, throughout the whole campaign from its inception in the Ellison Reef case. I make particular acknowledgement to *The Australian*, whose full and faithful coverage was a most important fact in the amount of public interest and information on the whole question.
>
> (Wright, 1977, p. xxiii)

In an early analysis of the campaign, Des Connell, then a chemistry PhD student and president of the Queensland Littoral Society, summarized for conservationists the lessons of the campaign and media roles:

> It is quite clear that sound and detailed knowledge of the problems and surrounding circumstances was a first essential requirement. This information then needs to be disseminated to the community by means of publications, public meetings, addresses to private organizations, and all the forms of mass media that can be utilized. Even after all of these actions have been taken, results cannot be achieved overnight, and both patience and perseverance are required.
>
> (Connell, 1971, p. 254)

While noting the impact of the press and television overall, Wright also acknowledged the impact of individual journalists. Barry Wain, at the then newly established Brisbane office of Rupert Murdoch's flagship *The Australian*, was singled out (Wright, 1977). Other journalists also figured in contemporary accounts, in particular Patricia Clare, who wrote the book *The Struggle for the Great Barrier Reef* (1971) based on interviews with vocal stakeholders, including scientists, farmers, activists, and local Indigenous people.

Media coverage of the 1969 Santa Barbara oil blowout in the United States also aided the campaign by raising "what if" anxieties in the Australian community. Concerns were not alleviated by comments from then Queensland State Mines Minister R.E. (Ron) Camm, who stated that similar damage could not occur because ocean currents would direct the oil away from the Reef. With 80% of the Reef leased for oil drilling permits, alarm was further amplified by Camm's comments that research could be conducted by pouring oil onto the Reef to determine its effect. In September 1969, *The Courier Mail* published a map of the reef region leased for oil drilling (in Wright, 1977, p. 71), naming the companies with leases – details that had been shrouded in secrecy until then. Federal politicians who had been at odds with the conservative Queensland state government supported a parliamentary enquiry that eventually led to passage of the Great Barrier Reef Marine Park Act. It defined Commonwealth rather than state jurisdiction over management of the Reef. The publication of that material also led to the revelation that Queensland Premier Joh Bjelke-Petersen held pecuniary interests in one of the mining companies and, thus, had a conflict of interest.

Initially, science and protests were on the margins of the debate. Mainstream media and individual sympathetic journalists brought scientific knowledge and contesting elite opinions to the center of public discourse. Büsst, Wright, Clouston, and key journalists developed a sophisticated media strategy. The campaign also used media to garner international support for the boycott of Ampol, which financially backed a Japanese company that was about to commence the first drilling at Repulse Bay. The Gold Coast branch of the Save the Reef Committee funded a full-page advertisement in the national newspaper, *The Australian*. That boycott eventually garnered support from trade unions at the peak of the controversy (Bowen & Bowen, 2002, p. 262; Wright, 1977, pp. 80–81). In 1968, the Save the Reef campaign presented the Federal Parliament with a petition of 13,000 signatures collected at public gatherings, including at fairs and airports.

The campaign also was aided by the existing social relations of the main protagonists, one of whom was Büsst's old university friend, Prime Minister Harold Holt. This gave the campaign direct access to the Parliament and a host of powerful national and international networks. The impact of these relationships should not be understated, even if it remains somewhat invisible to contemporary research. Policy development happened alongside, if not because of, the efforts of the Save the Reef campaigners. Nascent here was policy that would administer the Reef for conservation, including the Great Barrier Reef

Marine Park Act and the Reef's World Heritage listing, which would subject the Reef and Australian governments to an additional layer of surveillance and administration via the international policy program of the World Heritage Committee.

Throughout accounts of the campaign that emerged at the time, largely produced by "insiders," the role of media and communications was acknowledged as important to the campaign's success, but of secondary interest to the machinations of the protest movement and strategies, relationships, conflicts, power struggles, and policy outcomes. Media are presented to some extent as outside the protests, with campaigners feeding coverage and representation of their views to mainstream media and sympathetic journalists. Nevertheless, this chapter argues that Save the Reef had a clear focus on coverage throughout, and this publicity was vital to pressure governments, unions, lawyers, scientists, and the broader public to act.

Fight for the Reef

The context for the Fight for the Reef campaign is both familiar and distinctive from that which had framed the Save the Reef campaign. In the nearly 50 years between the campaign periods, global environmental consciousness, capital, and communications became entrenched. Global and local environmental policies designed to administer and protect the Reef from destruction also proliferated. In this context, Fight for the Reef's communication strategy is aimed at transcending national boundaries and bypassing traditional channels of power associated with media, government, and industry. It produces its own images, videos, and news that are widely distributed via its website, YouTube, and Twitter, and places broadcast advertising on commercial television. While many of these techniques, including aiming for international reach and advertising in mainstream media, were evident in the earlier campaign, the scale of this communication has multiplied exponentially, to use Cottle's (2014) criteria, with professional international environmental groups and more localized ones able to access audiences in the hundreds of thousands. To quantify, in comparison to the 13,000 signatures delivered to the Parliament in 1968, the Fight for the Reef Facebook site had over 170,000 likes, and its online petition with 81,000 signatures was delivered to the World Heritage Committee meeting in Cambodia in 2013. Greenpeace has delivered petitions containing hundreds of thousands of signatures to key World Heritage Committee meetings. Petitions calling on international financiers to divest from projects that support the Carmichael Mine project and other mining projects in the Galilee Basin have also been collected internationally (Rainforest Action Network, 2015). These figures do not consider the "network of networks" characteristic of social media where WWF Australia, WWF International, and other environmental NGOs tag and share online content.

Fight for the Reef's communication speed is also vastly faster than the earlier campaign's, with time and space collapsed to enable outcomes that carry the impact of suddenness. This is particularly the case with "Twitter storms" designed to coincide with World Heritage Committee meetings where Reef discussions took place. Different groups and memberships intersect in this new communications space, enabling social relations to connect and collate, thus empowering individuals and communities to share and support each other in the broadly united theme of mitigating disaster. Fight for the Reef formed alliances with other online protest movements including GetUp! Action for Australia, which boasts over 980,000 members and has been a tour de force in campaigning for social, economic, and environmental justice (see Forde, 2011). GetUp!, in partnership with local community organization Mackay Conservation Group and lawyers from the Environmental Defenders Office NSW, succeeded in having Adani's Carmichael Mine approval overturned in the Australian Federal Court in 2015. The communication network collated funding for a legal case, whose impact was felt across the local/global continuum. The federal government's counter was to attempt to limit legislation that empowers environmental NGOs to litigate against development projects where environments and biodiversity are threatened. Konkes (2018) asserts that the collision of public interest litigation, media, and environmental protest has received little academic attention and yet has been critical to the Reef, its conservation, and environmental conflict more broadly. Certainly, the campaign has enlisted media and law in ways that were impossible in 1967. In the later protest, emergent and established environmental law and policy provided a basis for protest and strategic disruptions of development proposals.

Local community knowledge and action intersect with international environmental and other NGOs, lawyers, and scientists with speed and impact. In this communications and media landscape, Fight for the Reef relies less on traditional media for the visibility of its environmental protest and conflict and writes its own news shared across networks. While recognition in mainstream media is still an ambition to influence political decision-makers and the implementation of policy, it is not critical to mobilizing public support and participation. Public mobilization is a process that in itself maintains the capacity to pressure those who make decisions.

Fight for the Reef has orchestrated strategic interventions and interruptions to the taken-for-granted authority of governments and the resources industry in particular via its networks. Mainstream politics and media are very much a part of this 21st-century communication milieu, working alongside these new media involving "simultaneously politics of representation and a politics of connectivity – both necessarily involved in scaling up and scaling down local global concerns" (Lester & Cottle, 2015, p. 108). This is evidenced by repeated high-profile and internationally distributed interventions in the Reef campaign by the Guardian news group, the involvement of Unilever's Ben and Jerry ice cream corporation via its "Scoop Ice Cream, not the Reef" campaign, and former US President Barack Obama's plea to the Australian government to ensure his daughters and their descendants would still be able to visit the Reef (Lester, 2015).

The impact of celebrity is also distinctive, symptomatic of a global communications industry where film stars, models, and other entertainment personalities are deemed newsworthy and powerful as a direct consequence of their participation in the entertainment industry (Lester, 2006; Boykoff & Goodman, 2009). In the case of Fight for the Reef, the impact of celebrity has been significant, especially in increasing visibility. For example, when actor Leonardo DiCaprio attended an Oceans Conference in Washington, DC, his comments on the Reef were reported widely via news services, tweeted, and shared on Facebook, including in *The Australian*:

> "Since my very first dive in the Great Barrier Reef in Australia 20-years-ago to the dive I got to do in the very same location just two years ago, I've witnessed environmental devastation first-hand," said DiCaprio, who pledged his own foundation would pump another $7 million over the next two years into projects to help the oceans.
>
> What once had looked like an endless underwater utopia is now riddled with bleached coral reefs and massive dead zones.
>
> "Unfortunately today, there's no proper law enforcement capacity and little accountability for violating the law. It's the Wild West on the high seas," warned the star of the box-office hit movie Titanic.
>
> (*The Australian*, 2014)

The campaign's response was to capitalize on DiCaprio's celebrity, naming him as a "Reef Champion" on its "Celebrity Wall." Governments and the Reef tourism industry were subsequently obligated to respond and have continued to defend the health of the Reef. In 2019, actor Simon Baker joined the campaign, uploading a YouTube clip on behalf of the Australian Marine Conservation Society and calling current government action to approve the Adani mine "reckless and terrifying" (Fight for Our Reef, 2019). The involvement of celebrity can create a "powerful environmentalism," although conservation messages do not always result in positive outcomes, especially when global perspectives marginalize or ignore local people and places (Brockington, 2008). The same powers that would threaten the Reef with industrial development may also bestow DiCaprio and other celebrities with their power in the public domain, arguably a contradiction of celebrity environmentalism and capitalism.

Governments and industry can use new media in a counter-strategy to negate the impact of dissent emerging from environmental protest. Moreover, the availability of media and its democratic function of giving a voice to citizens can result in a cacophony of voices. In the public sphere, this noise risks a redundancy of voices and issues, of protest or issue fatigue, as disasters, dangers, and threats compete for popularity or recognition. With these critical caveats, Fight for the Reef can produce its own TV commercials, broadcast on corporate Australian TV, and circulate its own maps to traditional media outlets but also via its network – and the network of networks. In ways both familiar and distinctive, this new media landscape maintains the democratic function of media, enabling citizen participation and voice in policy deliberations and decision-making, while also revealing still-emergent limitations to sustained influence and political power.

While the Fight for the Reef protest has sustained its media presence, another campaign has been at the forefront of recent Australian and international media attention. The #StopAdani campaign is part of a global alliance of organizations targeting the coal industry (Murphy, 2017). It has used media to identify and malign the public profile of international financial and insurance supporters of Adani and its Bravus Mining and Resources company as part of the Reef protest. Lester (2019) observed that international flows of finance, trade, politics, and media bring global attention to local places – flows susceptible to both local protest and international scrutiny. The 2019 federal election saw a #StopAdani convoy travel from the southern states of Australia to tell Queenslanders to choose corals over coal. While these efforts may have been misplaced (Tranter & Foxwell-Norton, 2020), they did focus attention on contemporary Reef conflict and the ongoing fight for protection that continues to traverse local and global contexts.

Conclusion

As Cottle (2008, p. 855) indicates, there is a wealth of opportunities to distribute dissent, often challenging the existing orthodoxy:

> Transnational communication and delivery systems now facilitate the internationalization and globalization of some protests. . . . [T]oday's complex media ecology, comprising different and overlapping media formations, horizontal and vertical communication flows and new interactional capabilities, offer unprecedented opportunities for the wider dissemination of political protest and dissent – from the local to the global.

This comparison of the protest campaigns Save the Reef and Fight for the Reef has revealed salient and distinctive changes to media and communications practices and flows across the 50-year period that the campaigns bookmark. Seemingly constant is the capacity of environmental protest, regardless of the era, to use media to pressure formal structures of environmental policy and legislation to impact perceived threats to the Reef from development applications. The most significant distinction between Save the Reef and Fight for the Reef is that political actors now recognize that expanded communication opportunities are more than mere vehicles for environmental protest. In the earlier Save the Reef era, campaigners represented themselves as passive, beholden to the whims of established networks of power. Fight for the Reef, on the other hand, represents itself as able to force dominant power structures to engage in interactions outside their control. In this way, new activist networks resemble their forebears, but with the qualification that societal power relations can be increasingly shaped and decided in the communications field (Castells, 2007, 2009).

Today, media roles in protests are acknowledged as active and seen to empower participatory politics that wrestle public debate from existing sites of cultural and economic power such as mainstream commercial media, industry, and governments.

How has the shift in communication strategies in the past 50 years affected outcomes for the Reef? The dominant motifs of both campaigns are the threats of industrial development and their potentially disastrous consequences. While the Reef itself currently is not to be mined, threats posed by the development of the resources industry remain, amplified within the context of climate crisis, a clear connection to fossil fuels and the impact of warming waters on its ecosystems. It might seem reasonable to have expected more for the Reef, given the perceived empowerment of activists via global communications and supported by interconnected layers of environmental policy and law – environmental policy and law that, in this case, are determined and administered across local, national, and international levels.

A community of dissent and a plethora of policy have not swayed the major Australian political parties and governments to discontinue their support for development proposals. The success of the Save the Reef campaign has been evident, with drilling halted, a raft of policies and laws established, and institutions like the Great Barrier Reef Marine Park Authority and the Australian Institute of Marine Science funded. Today's Fight for the Reef has achieved significant wins in impeding current development proposals but, despite its unparalleled access to communications tools and networks, has been unable to sway political decision-makers to abandon the proposals.

Opportunities continue for Great Barrier Reef protests to manufacture strategic interventions and interruptions or breakthrough moments that hold to account the powerful institutional actors of governments and industry – both national and international. Media and communications should present unprecedented power for civil society actors to affect formal processes of definitions and decision-making. Media democracy and ecological democracy intersect at the confluence of participation in media and in environmental decision-making (Foxwell-Norton, 2015, 2018). These critical and consequential processes strike to the heart of relations among media, politics, and environment in the 21st century. Yet, this analysis of the two campaigns crossing five decades, within the context of the current threatened state of the Reef, suggests that a power disjuncture remains between emerging media and communications practices and logics and enactment of sustained and meaningful policy. This power disjuncture is concerning, given that the modern era of the Great Barrier Reef takes place alongside the meteoric rise of policy, law, science, and institutions to protect the Reef and media capable of communicating and supporting conservation efforts (Foxwell-Norton & Konkes, 2019). And yet, today the Reef faces unprecedented anthropogenic threats that point to shifts in enduring power structures and relations among industry, party politics, science, environmental protest, and media (Hutchins & Lester, 2015; Konkes & Foxwell-Norton, 2020).

In the early 1960s, these forces collided. The eventual outcome was mitigation of plans to drill the Reef and establishment of institutions, law and policy, and an overall investment in Reef science. At this point in the 21st century, the legacy of these efforts to save the Great Barrier Reef is in doubt, as is the potential of a new media landscape – with all its breadth and capabilities.

Acknowledgments

This chapter draws on research previously published in Foxwell-Norton, K., & Lester, L. (2017). Saving the Great Barrier Reef from disaster, media then and now. *Media, Culture & Society*, *39*(4), 568–581. Used with permission from Sage Publications.

References

Adani. (2014). Carmichael coal mine and rail project. Retrieved from www.bravus.com.au/carmichael-mine/

The Australian. (2014). Leonardo DiCaprio tells of Great Barrier Reef "devastation." *The Australian*. Retrieved from www.theaustralian.com.au/national-affairs/leonardo-dicaprio-tells-of-great-barrier-reef-devastation/story-fn59niix-1226958336644

Australian Broadcasting Corporation. (2019). Adani gets final environmental approval for Adani mine. Retrieved from www.abc.net.au/news/2019-06-13/adani-carmichael-coal-mine-approved-water-management-galilee/11203208

Australian Government. (n.d.). *The World Heritage Convention*. Retrieved from www.environment.gov.au/heritage/about/world/world-heritage-convention

Beck, U. (2009). *World at risk*. Cambridge, UK: Polity Press.

Bowen, J., & Bowen, M. (2002). *The Great Barrier Reef: History, science, heritage*. Cambridge, UK: Cambridge University Press.

Boykoff, M. T., & Goodman, M. K. (2009). Conspicuous redemption? Reflections on the promises and perils of the "celebritization" of climate change. *Geoforum*, *40*(3), 395–406.

Brockington, D. (2008). Powerful environmentalisms: Conservation, celebrity, and capitalism. *Media, Culture and Society*, *30*, 551–568.

Carson, R. (1962). *Silent Spring*. Boston, MA: Houghton Mifflin.

Castells, M. (2007). Communication, power and counter-power in the network society. *International Journal of Communication*, *1*(1), 238–266.

Castells, M. (2009). *Communication power*. Oxford, UK, and New York: Oxford University Press.

Clare, P. (1971). *The struggle for the Great Barrier Reef*. Sydney: Collins.

Connell, D. W. (1971). The Great Barrier Reef conservation issue –a case history. *Biological Conservation*, *3*(4), 249–254.

Cosgrove, D. E. (1984). *Social formation and symbolic landscape*. London: Croom Helm.

Cottle, S. (2006). *Mediatized conflict*. Maidenhead, UK: Open University Press.

Cottle, S. (2008). Reporting demonstrations: The changing media politics of dissent. *Media, Culture and Society*, *28*(2), 853–872.

Cottle, S. (2014). Rethinking media and disasters in a global age: What's changed and why it matters. *Media, War, and Conflict*, 7(1), 3–22.

Cox, R. (2012). *Environmental communication and the public sphere* (3rd ed.). Thousand Oaks, CA: Sage.

Deloitte Access Economics. (2017). At what price? The economic, social, and icon value of the Great Barrier Reef. Retrieved from https://www2.deloitte.com/content/dam/Deloitte/au/Documents/Economics/deloitte-au-economics-great-barrier-reef-230617.pdf

Drew, P. (1994). *The coast dwellers: Australians living on the edge*. Victoria, Australia: Penguin.

Eagle, L., Hay, R., & Low, D. R.(2018). Competing and conflicting messages via online news media: Potential impacts of claims that the Great Barrier Reef is dying. *Ocean & Coastal Management, 158*, 154–163.

Evans, R. (2007). *A history of Queensland*. Melbourne: Cambridge University Press.

Fight for Our Reef. (2019). Acclaimed Australian actor Simon Baker joins the fight for our reef. Australian Marine Conservation Society. Retrieved from www.marineconservation.org.au/acclaimed-australian-actor-simon-baker-joins-the-fight-for-our-reef/

Forde, S. (2011). *Challenging the news: The journalism of alternative and independent media*. London: Palgrave Macmillan.

Foxwell-Norton, K. (2015). Community and alternative media: Prospects for 21st century environmental issues. In C. Atton (Ed.), *The Routledge companion to community and alternative media* (pp. 389–400). London: Routledge.

Foxwell-Norton, K. (2018). *Environmental communication and critical coastal policy: Communities, culture and nature*. London: Routledge.

Foxwell-Norton, K., & Konkes, C. (2019). The Great Barrier Reef: News media, policy, and the politics of protection. *International Communication Gazette, 81*(3), 211–234.

Foxwell-Norton, K., & Lester, L. (2017). Saving the Great Barrier Reef from disaster, media then and now. *Media, Culture and Society, 39*(4), 568–581.

Hansen, A. (2010). *Environment, media and communication*. New York: Routledge.

Hutchins, B., & Lester, L. (2015). Theorizing the enactment of mediatized environmental conflict. *International Communication Gazette*, 77(4), 337–358.

Hutton, D., & Connors, L. (1999). *A history of the Australian environment movement*. Melbourne: Cambridge University Press.

Konkes, C. (2018). Green lawfare: Environmental public interest litigation and mediatized environmental conflict. *Environmental Communication, 12*(2), 191–203.

Konkes, C., & Foxwell-Norton, K. (2020). Science communication and mediatised environmental conflict: A cautionary tale. *Public Understanding of Science, 16*(1).

Lester, L. (2006). Lost in the wilderness? Celebrity, protest and the news. *Journalism Studies*, 7(6), 907–921.

Lester, L. (2010). *Media and environment: Conflict, politics and the news*. Cambridge, UK: Polity Press.

Lester, L. (2014). Transnational publics and environmental conflict in the Asian century. *Media International Australia, 150*, 67–78.

Lester, L. (2015). Containment and reach: The changing ecology of environmental communication: News and new media roles. In A. Hansen & R. Cox (Eds.), *Routledge handbook of environmental communication* (Chapter 19). Abingdon, UK: Routledge.

Lester, L. (2019). *Global trade and mediatised environmental protest: The view from here*. Cham, Switzerland: Palgrave Macmillan.

Lester, L., & Cottle, S. (2009). Visualizing climate change: TV news and ecological citizenship. *International Journal of Communication, 3*, 920–936.

Lester, L., & Cottle, S. (2015). Transnational protest, publics and media participation (in an information age). In A. Hansen & R. Cox (Eds.), *The Routledge handbook of environment and communication* (pp. 100–110). London: Routledge.

Lester, L.,& Hutchins, B. (2009). Power games: Environmental protest, news media, and the internet. *Media, Culture & Society, 31*(4), 579–595.

Lester, L., & Hutchins, B. (Eds.). (2013). *Environmental conflict and the media*. New York: Peter Lang.

McCalman, I. (2013). *The Reef: A passionate history from cook to climate change*. New York: Viking, Penguin Books.

McGaurr, L.,& Lester, L. (2018). See it before it's too late? Last-chance travel lists and climate change. In B. Brevini & J. Lewis (Eds.), *Climate change and the media* (pp. 123–140). New York: Peter Lang.

Murphy, K. (2017, May 27). Federal labor feels the heat over Adani, and coalition is sweating too. *The Guardian*. Retrieved from www.theguardian.com/environment/2017/may/27/federal-labor-feels-the-heat-over-adani-and-coalition-is-sweating-too

NASA. (2014). *Apollo 8: Christmas at the moon*. Retrieved from www.nasa.gov/topics/history/features/apollo_8.html

Queensland Heritage Register. (n.d.). Ninney Rise and John Büsst Memorial. Retrieved from https://apps.des.qld.gov.au/heritage-register/explorer/detail/?id=602499

Rainforest Action Network. (2015). *Over 670,000 call on ex-im bank to reject controversial Australian coal project*. Retrieved from www.ran.org/press-releases/over_670_000_call_on_ex_im_bank_to_reject_controversial_australian_coal_project/

Tranter, B., & Foxwell-Norton, K. (2020). Only in Queensland? Coal mines and voting in the 2019 Australian federal election. *Environmental Sociology*. doi:10.1080/23251042.2020.1810376

Wright, J. (1977). *The coral battleground*. North Melbourne, Australia: Spinifex Press.

13 Communicating shark extinction

Celebrity-endorsed conservation campaigns and public policy in China

Elaine Jeffreys

The jury remains out on whether celebrity-endorsed environmental communication has a demonstrated positive impact in creating interested publics and transforming public policy agendas (Duthie, Veríssim, Keane, & Knight, 2017). Supporters argue that leveraging "fame" raises public awareness of environmental issues and organizations, as well as highlighting the need for legislative change by generating media publicity, clarifying campaign issues, and attracting new audiences and donors (https://wildaid.org). Critics counter that celebrity-fronted environmentalism is a self-serving celebrity branding exercise that oversimplifies issues by providing "spectacular enviro-tainment" (Goodman, 2013; Goodman, Littler, Brockington, & Boykoff, 2016). They assert that celebrity endorsement turns such issues into sensationalist media spectacles, matters of fan identification, and individual consumer choices ("causumerism"), rather than treating them as "big" political issues.

Discussions of the efficacy of celebrity-endorsed environmental communication typically focus on Western contexts with democratic governments and publics in mind. That invites the question of whether such communication can influence public policy elsewhere such as Asian authoritarian government contexts like the People's Republic of China (PRC), whose "huge environmental challenges are significant for us all" (Shapiro, 2012, p. 1).

WildAid, an international conservation organization, has used celebrities such as Shanghai-born Yao Ming, formerly a star player with the North American National Basketball Association, as the public face of communication aimed at ending the consumption of shark-fin soup in Chinese restaurants globally (https://wildaid.org). The soup is a luxury item consumed as a social event largely by ethnically Chinese people at banquets marking major occasions such as Chinese New Year and weddings, and at government and corporate events. Originally served as a tribute to the Chinese emperor, shark-fin soup honored the emperor because of the risk and expense involved in catching a "fierce" wild animal and a belief that some of the animal's vitality gets conferred on those who eat its fins (Jeffreys, 2016a, p. 763; see also Jarvis, 2019). Today, serving shark-fin soup honors a host and guests by confirming the host's wealth and implying that sponsors paying for the banquet respect the guests (see Figure 13.1).

Figure 13.1 Bowl of shark fin soup.

Source: From US Fish & Wildlife Service.

WildAid (2018, p. 4) claims that despite law enforcement efforts (see Figure 13.2), 100 million sharks are killed annually, mainly for the shark-fin trade, but also for the growing shark-meat market. Overfishing threatens to destroy the balance of ocean marine life by placing multiple shark species at risk of extinction. Some are de-finned alive and then thrown back into the ocean to die because the fins have more commercial value than the rest of the shark. Yet in spite of this value, shark fins reportedly have no flavor and contain toxins.

Using the now-standard marketing practices of international NGOs, WildAid has enlisted celebrity spokespeople to generate media publicity for its Shark Conservation Program, which advocates legislation to restrict shark fishing and tries to dissuade consumers from buying shark-fin products.

WildAid's shark conservation program is a translocal media strategy to connect and influence people around the world through the exchange of celebrity-endorsed information, ideas, and identities. The program communicates its core conservation message – "when the buying stops, the killing can too" – via multilingual broadcasts and social media. The celebrities who deliver this message have international, national, and local appeal.

The roll-out of specific campaigns mobilizes diverse networks of actors to transform government practices and individual behaviors, and it elicits different forms of sociability and community activity. These include expressions of cosmopolitan environmental awareness displayed through activism for anti-finning

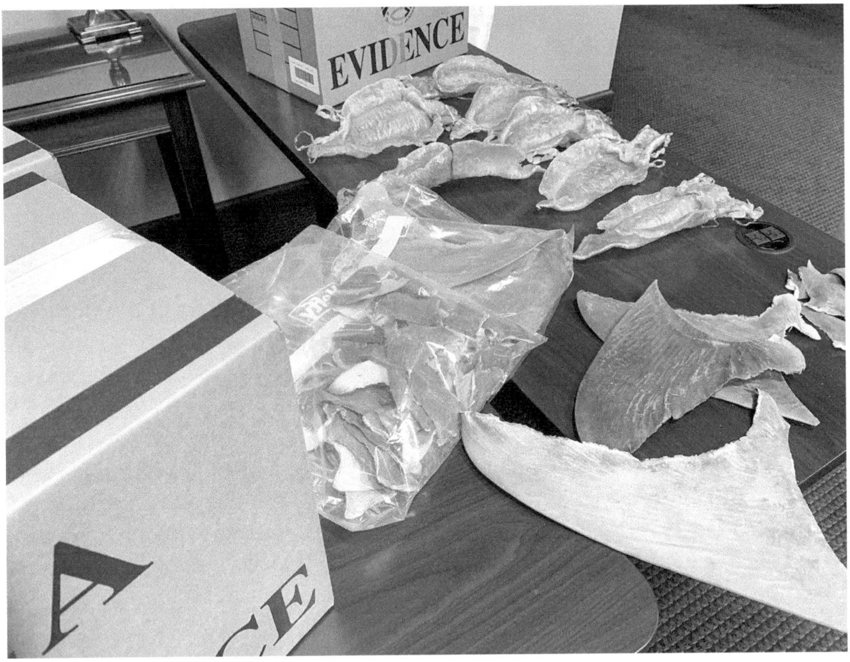

Figure 13.2 Sharks fins confiscated by US law enforcement agencies.

Source: From US Fish & Wildlife Service.

legislation and refusal to eat shark-fin dishes. However, other sentiments evoked by these campaigns are neither intended nor desirable. They include claims to the effect that "all Chinese go out of their way to eat endangered species because of superstitious health beliefs," and "all anti-shark finning activists are racist China bashers" (Jeffreys, 2016a, p. 764).

WildAid's shark-protection campaign illustrates philosopher-linguist George Lakoff's claim that effective communication about new ideas of, and action on, the environment needs "coherent framing." Why does it matter how we "frame" the environment? Lakoff (2010) argues that people do not comprehend the meaning of "facts" directly; rather, they think via mostly unconscious (but socially variable) interpretive contexts he calls "frames." Telling people "the facts" does not mean they will necessarily reach the "right conclusion" (pp. 72–73). Facts "will be ignored," unless they make sense in terms of culturally embedded systems of frames, which involve local institutions, industries, and practices (social roles, values, etiquette, etc.) (pp. 71–73, 77). To be communicated successfully, "the facts" must be framed persuasively and coherently, without activating related counterarguments and established habits (pp. 72, 79). Persuasive framing could involve using exemplary messengers, appropriate visuals, narratives that link frames and the message, framing issues in morally acceptable terms, and *never* repeating opposing arguments (pp. 79–80).

This chapter uses the concept of coherent framing to assess the efficacy of celebrity-fronted shark conservation campaigns in mainland China vis-à-vis "authoritarian environmentalism," a non-participatory approach to environmental policy formation and implementation (Gilley, 2012, p. 287). China is a focus of debate about the relative merits of "democratic versus authoritarian environmentalism" because of its comparatively rapid advances in government and public action on environmental concerns (Eaton & Kostka, 2014; Gilley, 2012).

Although the PRC's system of one-party rule ensured that the country had no independent NGO sector and ignored massive environmental degradation until recently, the government has the perceived capacity to implement environmental protection policies quickly. That is because, unlike Western governments, it does not need to accommodate the interests of high-consuming citizens as voters and can arguably ignore opposing business interests (Gilley, 2012, p. 288). A reported 80% decline in shark-fin imports to mainland China during 2012–2015 highlights this capacity. The decline was driven in part by NGO environmental communication, but also by top-down government austerity measures that resulted in a ban on shark-fin soup at state-sponsored banquets (Food and Agricultural Organization, 2019a, p. 15; WildAid, 2018, p. 4). Thus, the example of shark conservation and China underscores Lakoff's lament that environmental action tends to focus on individual behavioral change rather than political action, when "[to] an enormous degree, governmental action outweighs and shapes individual actions" (2010, p. 77).

Statement of the problem

Why do we need effective environmental communication about shark conservation? The message that must be communicated is clear: a quarter of shark species are threatened with extinction due to overfishing, primarily for the shark-fin trade, and the number of vulnerable species is likely to grow (Sadovy de Mitcheson et al., 2018, p. 116). This threat is underscored by the fact that an increasing number of species have been listed on the Convention on International Trade in Endangered Species of Wild Fauna and Flora (CITES), Appendix II (Cardeñosa et al., 2018). CITES is an international agreement among governments that aims to ensure that trade in specimens of wild animals and plants does not threaten their survival.

The shark trade as a whole is difficult to control because of limited knowledge of the increasingly globalized market in shark products (meat and fins), problems with trade reporting mechanisms, and limited law enforcement capability. Controls on the shark-fin trade are further complicated by the large number of sharks caught as secondary catch, high prices created by consumer demand, and difficulties in identifying fins by species (Dent & Clarke, 2015, p. iv; Sadovy de Mitcheson et al., 2018, p. 118). Despite a rising number of countries' bans on shark finning, anti-finning legislation typically bans the practice of catching a shark solely for the purpose of harvesting its fins. It is illegal to import and export fins that are not attached to a shark; however, fins obtained from managed shark fisheries and unprotected species can still be traded (Animal Welfare Institute, 2020; Guida, 2019; Schiffman et al., 2020, p. 2). The emergence of new markets for shark meat also means that pressure to improve international and national monitoring and regulatory systems must continue (Dent & Clarke, 2015, pp. 1–2; WildAid, 2018). Hence, for the foreseeable future, conservation communication is likely to remain an integral part of efforts to reduce consumer demand and to attract support for policy changes to put species and target controls on shark fishing, and/or to enforce existing bans.

Who are the world's largest importers of shark fins? The Food and Agricultural Organization of the United Nations (FAO) provides the most up-to-date and comprehensive trade volume data. According to the FishStat Database (Global Fisheries Commodities Production and Trade), the five largest importers, accounting for around 95% of average annual global imports during 2000–2017, were Hong Kong, Malaysia, China, Singapore, and Taiwan. However, China has dropped off the list, with Canada coming in at fifth place based on the 2011–2017 figures (2019b).

The global market has been gauged historically with reference to statistics from Hong Kong, a major shark-fin trade hub (Cardeñosa et al., 2018; Shea & To, 2017). Imports into Hong Kong halved between 2007 and 2017; the re-export of shark fins declined by three-quarters over the same period (WWF, n.d.). PRC government reports similarly record a sharp decline in imports over 2012–2015 (Hofford, 2017; Shea & To, 2017, p. 333).

There are multiple reasons for the decline in the Hong Kong trade, and especially exports to China, not all of which bode well for shark survival (Dent & Clarke, 2015, pp. 6, 26, 36–38; Eriksson & Clarke, 2015, p. 169; WildAid, 2018, p. 4). One factor is shifting trade dynamics due to the introduction of taxes on shark-fin imports to mainland China from Hong Kong following the PRC's entry to the World Trade Organization in 2001. That increased the cost of processing shark fins in China for re-export. Shifting and emerging new markets, including increased domestic shark-meat production in China, and perhaps even reduced supplies of fins due to overfishing, also played a role. Other factors include growing conservation awareness, consumer backlash against artificial shark-fin products, increased regulation of finning and other trade bans (retailer and transportation boycotts), and curbs on government officials' expenditures in mainland China. Statistical measurement changes may also have contributed to the reported decline. Changes to Hong Kong trade classifications in 2012–2014 resulted in frozen fins being re-classified as frozen shark meat, and there may have been underreporting of some trade. Although scientists hypothesize that environmental communication has not played a major role in the decline in the Hong Kong shark-fin trade to date, they acknowledge that it is impossible to determine the relative importance of each of these factors (Eriksson & Clarke, 2015, p. 169).

Why was the mainland Chinese market for fins further suppressed after 2012? The two commonly cited reasons are: 1) celebrity-endorsed environmental communication by international conservation organizations educating people about the need to protect sharks and 2) introduction of government anti-corruption and austerity measures in late 2012, accompanied by bans on consumption of high-end cuisine at official banquets, including dishes that contain shark fin (Jeffreys, 2016a; WildAid, 2018). International organizations such as WildAid and the World Wildlife Fund (WWF) naturally attribute this drop to the success of their environmental communication, and only secondarily to the government ban (Kao, 2014; Jin shi yuchi, 2012; Yuan, 2012; WildAid, 2018). However, as the next section shows, government-led policy formation may also have played a major role.

Methods

This chapter uses an illustrative case study, drawing on statistical data and media content analysis, to reveal the relationship between celebrity-endorsed shark protection campaigns and PRC government regulations exacerbating a decline in the demand for shark fins in China. Statistical data on annual shark-fin trade volumes in mainland China came from the FAO's FishStat Database.

Media analysis is a recognized way of observing environmental communication and related government, industry, NGO, and social networks (Boykoff & Boykoff, 2007). The author accessed English-language media content by searching for combinations of the keyword "shark fin" with, respectively, "Yao Ming," "WildAid," "public money," and "corruption" (four combinations in total), after 2005 on Factiva.com an international news database produced by Dow Jones. Yao Ming signed a WildAid pledge not to eat shark fins in 2006 (Jeffreys, 2016a, p. 766). The author accessed Chinese-language media content by searching for the same four combinations of the equivalent Chinese-language keywords (鱼翅 + 姚明, 野生救援, 公款, 腐败) on the China Core Newspaper Full-text Database of the China Knowledge Resource Integrated Database (CNKI) after 2005. The study supplemented broadcast media content with searches for related content on Sina Weibo, a microblogging website and one of China's biggest social media platforms. However, the chapter emphasizes broadcast media content because only 10% of the Chinese population had internet access in 2006, and only half of the population had internet access in 2015 (Internet Live Stats, 2016). In comparing international NGO marketing and Chinese media content, the case study discusses contemporary mediatized environmental communication as a product of the often-unpredictable interactions among media, industry, NGO activism, and formal politics.

Findings

Table 13.1 shows change over time (2000–2017) in China's trade with the rest of the world in dried, prepared, and preserved shark fins, sourced from

Table 13.1 China – shark-fin trade: percentage change in net product weight: sum of dried, prepared, and preserved shark fins.

	Exports	*Imports*
2000–2005	– 32.2	– 27.8
2006–2011	– 14.1	– 94.0
2012–2017	– 16.5	– 59.3

Sources: Author's calculations; Food and Agriculture Organization (2019a, 2019b).

the FishStat Database (Global Fisheries Commodities Production and Trade) (FAO, 2019b). Data on trade in fresh, chilled, and frozen fins are excluded due to apparent gaps in reporting. Overall, the data indicate that both exports and imports experienced a major decline beginning in the early 2000s, before the start of WildAid's China-focused shark conservation campaigns. However, imports declined much faster than exports, suggesting that China switched from being a net importer of shark-fin products to a net exporter during the 2010s. Numerous reasons may account for the dramatic (97%) decline in imports of shark fins to China from 2000 to 2011, including growing conservation awareness. However, it is striking that even after imports had already fallen to relatively low levels by the standards of the early 2000s from more than 4,000 metric tons (4 million kg) per annum to less than 200 metric tons (200,000 kg), they experienced a further 59% decline after 2011. It is plausible that bans on state-sponsored consumption of shark-fin dishes, which were announced in mid-2012 and introduced in late 2013, intensified the pre-existing downward trend in Chinese imports.

Although the decline in imports could be interpreted as also indicating expansion of the Chinese domestic industry, the available evidence suggests otherwise. Despite the lack of official statistics on Chinese shark-fin production, the FAO's FishStat Database (Regional Capture Fisheries Statistics) sheds some light on Chinese shark capture numbers. For the seven species with continuous capture figures available since 2007, the data shows a pronounced decline in production in the late 2000s and relatively stable production since then. While not conclusive, the figures point to a decline in domestic consumption (rather than an expansion of domestic production) as the major factor behind the fall in imports. This inference is supported by a consumer behavior survey commissioned by WildAid, which reports a significant decline (more than 80%) in cities across China in sales by vendors and restaurants, and in people eating shark fin, partly due to concerns about fake products (Whitcraft et al., 2014).

The available evidence suggests that consumer demand remains low in China, even though shark conservation communication has diminished. WildAid (2011) achieved significant media exposure for its conservation communication. For example, it obtained over $200 million in pro-bono media support and reached more than 1 billion people in China in 2011. However, the media response to WildAid's conservation communication was greater in international arenas than in China. A search for keywords "shark fin" and "Yao Ming," and "shark fin" and "WildAid," on Factiva.com between 2010 and 2012 – the height of the Yao Ming shark conservation campaign – produced 240 results in total, while a Chinese-language search on the CNKI Database found 28 results for the same period, with numbers dropping dramatically afterwards (cnki.net). By comparison, a search for the keywords "shark fin" and "public money," and "shark fin" and "corruption," on Factiva between 2012 and 2014 – the period when consumption became a focus of formal policy making and implementation in China – produced around 140 hits, while a Chinese-language search on the CNKI Database found more than 330 results for the same period. Searches on Sina Weibo further reveal that Chinese social media discussions of sharks were a hot-button issue during 2012–2014, but there has been limited discussion since then (weibo.com).

The CNKI figures point to the greater prominence of government action on consumption in mainland China compared with actions by international conservation organizations. Unsurprisingly, the PRC government not only exercises considerable control over Chinese media content, but also decides which social and environmental problems should be fixed (or not). However, the decline of the shark-fin trade to current low levels suggests that a better understanding of the PRC case may offer insights for future environmental communication.

The next section illustrates these points by detailing the two interventions that plausibly help explain declining demand in China: first, WildAid's media campaign, and second, the PRC government-led austerity campaign.

Discussion and conclusions

WildAid's environmental communication, which targeted China as the world's historically leading consumer market for shark fin, adopted a key principle of coherent framing – tell a story that "rouses emotions" while overlooking injunctions not to "repeat any counter arguments" (Lakoff, 2010, p. 79). In 2006, WildAid launched a series of celebrity-fronted public service advertisements (PSAs) in China promoting protection of endangered species. They aired without charge on cable networks and the China Central Television station (Jeffreys, 2016a). This communication capitalized on the then unusual international attention centered on China before the 2008 Beijing Olympic Games.

One of the earliest sets of China-focused shark PSAs featured three female Olympian divers who appeared in short, visually spectacular videos in English and Chinese (Jeffreys, 2016a, p. 766). The PSAs communicated that: (1) sharks are a vital part of the ocean ecosystem and should be protected; (2) millions are killed annually because serving shark-fin soup at banquets honors the guests and host, or gives them "face"; (3) sharks are being decimated; and (4) China and Chinese people will "lose face" when that happens, so "don't eat shark-fin soup!" They concluded with the WildAid tagline: "When the buying stops, the killing can too." The narratives thus combined environmental communication with appeals to national pride, suggesting that China's international reputation is more important than the forms of sociability associated with eating imperial cuisine. But the PSAs also demonstrate incoherent framing by potentially triggering the notion that eating shark-fin soup is a custom that "gives face" to individuals in more localized contexts.

Banquet behavior similarly framed Yao Ming's first PSA as a solo spokesperson on shark conservation. The PSA distributed before the 2010 Chinese New Year takes place in an upmarket restaurant with a wall-to-wall aquarium. It states, "What if you could see how shark-fin soup is made?" As a waiter carries bowls of soup to a table, the viewer sees blood pouring from the cut-off fins of a still-alive shark lying at the bottom of the aquarium (Shipin: Yao Ming, n.d.; Yao Ming – shark-fin soup, 2009). A narrator says, "If you could see how each year up to 70 million sharks are killed to end up in soup, could you still eat it?" The camera shows ethnically Chinese diners looking at the shark. Yao Ming pushes away a bowl of soup. The narrator continues, "A third of all shark species are nearly extinct, but we can help save them." As other diners begin to push away their bowls, Yao Ming concludes, "Remember, when the buying stops, the killing can too."

This PSA assumes that empathy with an edible animal supersedes host–guest relations and respect for tradition/custom. It ignores the possibility that refusing the soup might insult the host, and hence may damage continued family and/or work relationships. What makes the framing incoherent is that the invocation of tradition/custom creates an inconsistency between the tagline and the spectacle. The tagline equates "*buying*" to "*killing*," but the spectacle goes much further, more contentiously equating "*accepting hospitality*" to "*killing*."

A second Yao Ming PSA, released in 2011, opens with him looking over a sunny San Francisco Bay and asking: "What is the price of shark-fin soup? If we keep killing tens of millions of sharks each year just for soup, it will change life in our oceans forever" (Yao Ming – the price of shark-fin soup, 2011). The camera then reveals an apocalyptic image of a dry landlocked basin that has replaced the saltwater bay. Yao continues: "Is this the world we want to leave our children? . . . Let's keep sharks in our oceans, not in our soup." He concludes with the WildAid tagline, reminding viewers that ethical consumption can save the planet for future generations. The attention-getting spectacularism of this PSA potentially undermines the coherence of its environmental communication as there is no evident connection between eating shark fin and "dried-up" oceans.

WildAid's preference for spectacular messaging may explain the larger media take-up of its shark communication with international vis-à-vis domestic Chinese audiences. Journalists credit the Yao Ming PSAs for contributing to bans on shark fin in 2011 in California and Toronto – areas with large proportions of ethnically Chinese residents (Harmon, 2011; Rogers, 2013). Hailed as a victory against a "savage, barbaric practice" by conservation groups and pro-shark politicians, both bans faced court challenges for singling out a Chinese custom, rather than banning shark products in general (Rider, 2013; Rogers, 2013).

Although the Toronto ban was overturned, the California ban was upheld. In 2019, Canada banned the import and export of fins that are not attached to a shark (Animal Welfare Institute, 2020).

WildAid also engaged in extensive shark-protection collaborations with Chinese national-level corporate and entertainment celebrities, as illustrated by the release of a 2012 Chinese-language PSA featuring five influential CEOs (Shipin: Yesheng jiuyuan, 2012). The PSA engages coherent framing by linking environmental protection with issues of global risk and national pride, while arguably using the "wrong messengers" (Lakoff, 2010, p. 80). Business leaders are aspirational role models in China, but "distrust" of the newly rich is common (Jeffreys, 2016a, p. 768).

The PSA opens with the five CEOs sitting around a dining table, and asks: "When these business leaders come together, what do they talk about?" The CEOs say in turn: "We must plan for the future. We must protect our assets. We must think globally and act locally. We must make the world proud of China. We must protect our oceans by not eating shark fin." It concludes with the WildAid tagline, implying that Chinese people who seek to be affluent, patriotic, cosmopolitan, and acclaimed, are forward-thinking and not tradition-bound; they can save the world's oceans by choosing not to eat a dish once prepared for emperors.

The CEOs are members of the China Entrepreneur Club (CEC), an elite not-for-profit whose links with WildAid highlight the emergence in China of elite-led translocal networks of action focused on the environment. In 2008, the CEC and China's Ocean University released a "shark fin nutrition report" revealing that fins have no nutritional value and contain toxins (Jeffreys, 2016a, p. 769). In 2009, Zhang Xingsheng, a CEC member and director of the North Asia region of the Nature Conservancy, started a conversation on China's Twitter-like Weibo, calling on the PRC's legislature to ban shark-fin consumption. Zhang (2011) has over 1.6 million Weibo followers, but the shark conversation attracted fewer than 330 responses.

In 2009, the CEC celebrated Earth Day by launching a "Protect Sharks: Don't Eat Shark Fin" initiative with WildAid and domestic environmental groups (Shu baiming qiyejia xiangying, 2009). CEC members pledged to neither eat shark fin nor accept or give gifts of fin. Calls for pledges were posted on a dedicated website titled "Stop: No Shark-fin Buying, No Sharks Killing" (2009), receiving over 2,400 responses. Celebrity CEOs such as Wang Shi (2009) reposted messages about the pledge on their blogs – over 31 million people read Wang's blog, although his shark thread attracted fewer than 150 posts.

The CEC and other organizations also ran multiple shark-protection activities in 2011–2013, using the slogans: "Don't Eat Shark Fin [for 2012 New Year's Banquets]" and "Create Zero-Shark-Fin Cities: Don't Eat in Shark-Fin Restaurants" (Jeffreys, 2016a, pp. 769–770). Tencent, China's largest internet service portal, created a "Don't Eat Shark Fin" website, which listed affiliates, provided news updates, and featured the WildAid tagline (Juchi yuchi, n.d.).

But only 1,600 people had clicked the "I promise not to eat shark fin" pledge as of 2015. Around 100 organizations signed an unsuccessful petition urging the Shenzhen government to make Shenzhen China's first "SharkFin Free City."

Although these initiatives involved action at an unambiguously influential level, they failed to get visible support, as the organizers admitted in media interviews (Jeffreys, 2016a, p. 770). Support chiefly came from multinational hotel chains and domestic restaurant chains with international aspirations, which had already committed to supporting WildAid. The "Zero Shark Fin" organizers subsequently launched initiatives to attract volunteers in various cities, working on the assumption that 1,000 volunteers in each city would translate into broader action through social media networks. These efforts also met with limited success.

The failure of these efforts to attract visible public support suggests that the "wrong messengers" fronted the environmental communication (Lakoff, 2010, p. 80). This conclusion is partly confirmed by responses to a WildAid youth-focused campaign called "I'm Finished with Fins," which launched in 2013 with support from the celebrity hosts of *Happy Camp*, a hugely popular family variety show. The associated PSA on Youku, a Chinese version of YouTube, attracted more than 83,000 views and multiple posts on other social media totaling more than 360,000 views (Jeffreys, 2016a, p. 771; Shipin: Kuailejiazu wo yu yuchi, n.d.). Many posts reiterated WildAid's communication that "shark finning is brutal." However, others accused the celebrity hosts of being sanctimonious because "only rich people can eat such things." Yet others suggested that environmental communication should be posted in government departments – because government officials were known to be major consumers of luxury seafood items – rather than on sites used by "ordinary citizens" (Jeffreys, 2016a, p. 771).

However, WildAid's leveraging of elite networks succeeded where it might have been most expected to fail: It brought the issue of shark-fin consumption to the March 2012 meeting of the National People's Congress (NPC) – China's so-called "rubber stamp" legislature. The full congress of nearly 3,000 party-selected but publicly "elected" members meets for around two weeks every year to vote through bills, the main content of which is determined by party committees and government departments. While members, including major CEOs and entertainment celebrities, are expected to submit motions on alternative public policies, and increasingly do so, such motions are rarely adopted (Jeffreys, 2016b).

Ding Liguo, a CEC member, steel magnate, and NPC representative, submitted a motion drafted by Zhang Xingsheng from the Nature Conservancy recommending that government departments set an international example of good practice by banning shark-fin soup at state-sponsored banquets (Jeffreys, 2016a, p. 771). A publicized reply in mid-2012 from the State Council, the PRC's highest organ of state administration, stated that legislation would result by 2015. In late 2013, the Central Committee of the Communist Party of China and the State Council issued the Regulations on Domestic Official Hospitality for Party and Government Organs. Article 9 states that official dinners should not involve high-end cuisine or dishes that contain shark fins, bird's nests, or any protected animal. The sale of shark fins promptly further declined in mainland China.

But the primary government motivation for banning shark-fin soup at official banquets was political survival rather than environmentalism (Jeffreys, 2016a). Between late 2012 and early 2013, incoming President Xi Jinping announced measures to combat government corruption and extravagant spending on official receptions, stating that the persistence of such problems would lead to the downfall of the party and government (Xi, 2014, pp. 17–18, 399). The State Council presumably acted on Ding Liguo's proposal because the central government had already committed to the now-popular austerity measures.

While inadvertently highlighting the potential of authoritarian environmentalism, the greater media visibility of the government austerity measures relative to shark-protection initiatives may be explained by the different sources of demand for shark-fin dishes – government officials versus private individuals. WildAid's celebrity and elite-endorsed conservation communication was framed in terms of changing individual discretionary behavior, especially that of young, well-off Chinese. Shark protection campaigns clearly have the capacity to alter individual consumption patterns, as evidenced by the success of the Yao Ming environmental communication in North America. The authority of celebrity endorsements may also be persuading young Chinese adults that eating shark fins is bad for the environment.

But the PSAs' evocation of banqueting and associated social relations arguably involved incoherent framing in the Chinese context as it misrepresented the choices available, culturally, to recipients of hospitality. Shark-fin consumption in China is rarely framed at an individual level; the main purchasers are also older than the target audience of celebrity-endorsed conservation communication. They are parents of young adults who are getting married with expensive celebratory banquets, or people with sufficient access to organizational resources to afford prestige-giving banquets. More generally, the framing may lack effectiveness because most young adults in China are unlikely to purchase expensive items such as shark fins in the first place.

Instead, the government austerity measures directly affected the behavior of elites whose luxury consumption was driven by the need to build intra-party, intra-government, and business relationships to achieve personal success, and by the availability of public funds. These measures had an almost instant effect, based on media reports of high-end restaurants closing or changing their menus to accommodate the new constraints on officials' hospitality expenditures (Dent & Clarke, 2015, p. 38; Jeffreys, 2016a, p. 772). This success illustrates how an environmental communication campaign aiming to transform individual consumer choices can become a political "big issue" once an authoritarian government links it to government spending.

The austerity measures reinforced and extended the publicity associated with the earlier celebrity-endorsed shark conservation campaigns. The influence of such conservation communication, which dovetailed with an emerging political agenda, spotlights the potential for environmental organizations and elite syndicates to effect social change in China. These latter organizations managed not only to persuade national and local elites to roll out shark protection campaigns across numerous cities (despite the general failure of such activities), but also brought shark conservation to the attention of China's legislature. The scale of that enterprise shows how NGO and elite-led communicative activism can be self-reinforcing, leading one group of organizations and elites to inspire other groups to undertake similar activities, regardless of national boundaries. This suggests that celebrity-corporate activism could be an important force for future action on the environment in China, given the limited spaces available for broader political participation in that country.

Although the decline in China's shark-fin trade cannot solely be attributed to celebrity-endorsed conservation communication and government austerity measures, both contributed to a reduced demand at elite, government-corporate banquets. This case study consequently highlights the complementarities between activism led by international conservation organizations and top-down government action in advancing environmental sustainability objectives. However, for international environmental activism to be more significantly successful in country-specific settings, the use of communication practice based on "Western" models clearly needs refining. It not only matters how we frame the environment and sharks, but also how we understand the environment of human action in different cultural contexts.

Acknowledgments

This chapter draws on research previously published in Jeffreys, E. (2016). Trans-local celebrity activism: Shark-protection campaigns in mainland China. *Environmental Communication, 10*(6), doi:10.1080/17524032.2016.1198822. It is published with the permission of Taylor & Francis Ltd., www.tandfonline.com.

References

Animal Welfare Institute. (2020). International shark finning bans and policies. Retrieved from https://awionline.org/content/international-shark-finning-bans-and-policies

Boykoff, M., & Boykoff, J. (2007). Climate change and journalistic norms: A case-study of US mass-media coverage. *Geoforum, 38*, 1190–1204.

Cardeñosa, D., Fields, A., Babcock, E., Zhang, H., Feldheim, K., Shea, S., . . . Chapman, D. (2018). Cites– listed sharks remain among the top species in the contemporary fin trade. *Conservation Letters*. doi:10.1111/conl.12457

Dent, F., & Clarke, S. (2015). *State of the global market for shark products*. Food and Agriculture Organization of the United Nations. Retrieved from www.fao.org/3/a-i4795e.pdf

Duthie, E., Veríssimo, D., Keane, A., & Knight, A. (2017). The effectiveness of celebrities in conservation marketing. *PLoS One, 12*(7), e0180027. doi:10.1371/journal.pone.0180027

Eaton, S., & Kostka, G. (2014). Authoritarian environmentalism undermined? Local leaders' time horizons and environmental policy implementation in China. *The China Quarterly, 218*, 359–380.

Eriksson, H., & Clarke, S. (2015). Chinese market responses to overexploitation of sharks and sea cucumbers. *Biological Conservation*, *184*, 163–173.

Food and Agricultural Organization. (2019a). *FAO yearbook: Fishery and aquaculture statistics 2017*. Retrieved from www.fao.org/fishery/static/Yearbook/YB2017_USBcard/booklet/CA5495T_web.pdf

Food and Agricultural Organization. (2019b). FishStat plus – universal software for fishery statistical time series. Retrieved from www.fao.org/fishery/statistics/software/fishstat/en

Gilley, B. (2012). Authoritarian environmentalism and China's response to climate change. *Environmental Politics*, *21*(2), 287–307.

Goodman, M. (2013). Celebritus politicus, neoliberal sustainabilities and the terrains of care. In G. Fridell & M. Konings (Eds.), *Age of icons: Exploring philanthrocapitalism in the contemporary world* (pp. 72–92). Toronto: University of Toronto Press.

Goodman, M., Littler, J., Brockington, D., & Boykoff, M. (2016). Spectacular environmentalisms: Media, knowledge, and the framing of ecological politics. *Environmental Communication*, *10*(6), 677–688.

Guida, L. (2019). If shark finning is illegal, how come Australia is still part of the shark fin trade? *Australian Marine Conservation Society*. Retrieved from www.marineconservation.org.au/shark-finning-is-illegal-but-australia-part-of-shark-fin-trade/

Harmon, S. (2011, October 7). Shark trade banned in California: Gov. Jerry Brown signs bill banning shark fins in California. *Mercury News*. Retrieved from www.mercurynews.com/2011/10/07/gov-jerry-brown-signs-bill-banning-shark-fins-in-california/

Hofford, A. (2017, January 6). Air China bans shark fin: Imports to China down 82% in 3 years. Retrieved from https://wildaid.org/air-china-bans-shark-fin-imports-to-china-down-82-in-3-years/

Internet Live Stats. (2016). China Internet users. Retrieved from www.internetlivestats.com/internet-users/china/

Jarvis, J. (2019). Shark fin soup: Collective imagination in the transnational public sphere. *Global Media Journal*, *11*(1), 49–63.

Jeffreys, E. (2016a). Translocal celebrity activism: Shark protection campaigns in mainland China. *Environmental Communication*, *10*(6), 763–776.

Jeffreys, E. (2016b). Political celebrities and elite politics in contemporary China. *China Information*, *30*(1), 58–80.

Jin shi yuchi: Lifa de shangjie liliang [Banning shark fin consumption: The legislative power of the business community]. (2012, July 27). *Lü Gongsi*. Retrieved from http://finance.eastmoney.com/news/1350,20120727242004096.html

Juchi yuchi [Say no to shark fin]. (n.d.). *Tencent*. Retrieved from http://news.qq.com/zt2011/bcyc/

Kao, E. (2014, April 8). Shark fin trade from Hong Kong to China drops almost 90 percent in one year. *South China Morning Post*. Retrieved fromwww.scmp.com/news/hong-kong/article/1469412/hong-kong-shark-fin-imports-fall-35pc?page=all

Lakoff, G. (2010). Why it matters how we frame the environment. *Environmental Communication*, *4*(1), 70–81.

Rider, D. (2013, February 19). Shark-fin ban resurfaces as Toronto council backers seek new, narrower rule. *The Star*. Retrieved from www.thestar.com/news/city_hall/2013/02/19/sharkfin_ban_resurfaces_as_toronto_council_backers_seek_new_narrower_rule.html

Rogers, P. (2013, July 1). Shark fin soup ban takes effect Monday. *Mercury News*. Retrieved from www.mercurynews.com/science/ci_23567375/shark-fin-soup-ban-takes-effect-monday

Sadovy de Mitcheson, Y., Andersson, A., Hofford, A., Law, C., Hau, L., & Pauly, D. (2018). Out of control means off the menu: The case for ceasing consumption of luxury products from highly vulnerable species when international trade cannot be adequately controlled; Shark fin as a case study. *Marine Policy*, *98*, 115–120.

Schiffman, D. S., Bittick, S. J., Cashion, M. S., Colla, S. R., Coristine, L. E., Derrick, D. H., . . . Dulvy, N. K. (2020). Inaccurate and biased global media coverage underlies public misunderstanding of shark conservation threats and solutions. *iScience*, *23*(6), 1–13.

Shapiro, J. (2012). *China's environmental challenges*. Cambridge: Polity Press.

Shea, K. H., & To, A. (2017). From boat to bowl: Patterns and dynamics of shark fin trade in Hong Kong – implications for monitoring and management. *Marine Policy*, *81*, 330–339.

Shipin: Kuailejiazu wo yu yuchi "Wo yu yuchi shuo zaijian" huodong xuanchuan pian [Video: Happy Camp "I'm Finished with Fins" promotion activity video]. (n.d.). *Youku*. Retrieved from http://v.youku.com/v_show/id_XNjEwNDg3ODY4.html

Shipin: Yao Ming: Buchi yuchi xuanchuan pian_yesheng jiuyuan_qiankewang [Video: Yao Ming: Don't eat shark fin PSA_WildAid_Qianke]. (n.d.). *Youku*. Retrieved from http://v.youku.com/v_show/id_XMjA1MjM2ODc2.html

Shipin: Yesheng jiuyuan gongyi guanggao [Video: WildAid PSA]. (2012). Retrieved from http://v.youku.com/v_show/id_XNTIzNDU3NzQ4.html

Shu baiming qiyejia xiangying "baohu shayu, juchi yuchi" gongyi xingdong [Hundreds of entrepreneurs join the "Say no to shark fin" action]. (2009, April 23). *Zhongguo Jingji Wang*. Retrieved fromwww.ce.cn/xwzx/gnsz/zg/200904/23/t20090423_18893604.shtml

Stop: No Shark-fin Buying, No Sharks Killing. (2009, April 22). *Zhongguo Qiyejia Julebu*. Retrieved fromwww.daonong.com/shark/shark_login.html

Wang, S. (2009, June 26). Juchi yuchi [Say no to shark fin], Shan zai na – Wang Shi blog. *Sina.com*. Retrieved fromhttp://blog.sina.com.cn/s/blog_47233d0f0100dkus.html

Whitcraft, S., Hofford, A., Hilton, P., O'Malley, M., Jaiteh, V., & Knights, P. (2014). Evidence of declines in shark fin demand, China. WildAid. Retrieved from https://wildaid.org/wp-content/uploads/2017/09/SharkReport_Evidence-of-Declines-in-Shark-Fin-Demand_China.pdf

WildAid. (2011). Annual report 2011. Retrieved from https://wildaid.org/wp-content/uploads/2017/10/2011-Annual-Report.pdf

WildAid. (2018). Sharks in crisis. Retrieved from https://wildaid.org/wp-content/uploads/2018/02/WildAid-Sharks-in-Crisis-2018.pdf

WWF. (n.d.). Appendix A – shark product import and re-export volume. Retrieved from https://d3q9070b7kewus.cloudfront.net/downloads/shark_product_import_and_re_export_volume_eng.pdf

Xi, J. P. (2014). *The governance of China*. Beijing: Foreign Languages Press.

Yao Ming – the price of shark-fin soup. (2011). *WildAid* (Video). Retrieved from www.youtube.com/watch?v=G7oaTCBuI2g

Yao Ming – shark fin soup. (2009). *WildAid* (Video). Retrieved fromwww.youtube.com/watch?v=mJG7RaLX-DM&feature=relmfu

Yuan, J. (2012, July 25). Jinshi yuchi: Lifa de shangjie liliang [Banning fin consumption: The legislative power of the business community]. *Lügongsi*. Retrieved from www.daonong.com/g/17/tianxia/2012/0725/34647.html

Zhang, X. (2011, January 21). Baohu daziran cong jujue yuchi kaishi [Protecting nature starts by refusing to eat shark fins]. *Weibo*. Retrieved from http://talk.weibo.com/ft/20110118248

14 Manatees and fossil-fuel power plants

The paradox of endangered species laws

John Carr and Tema Milstein

During central Florida's winter months, visitors swarm to Tampa Electric Company's (TECO) Manatee Viewing Center to catch sight of Florida manatees. In the cold months, these endangered animals live by necessity in the shadow of the company's Big Bend Power Station, a towering coal and hydrocarbon gas–burning power plant. It is not uncommon to see hundreds of manatees swimming in this vast industrial site's effluent stream, a space legally designated as their protective sanctuary. While it may seem paradoxical to designate a fossil-fuel-burning power plant as an endangered species' sanctuary, this study traces this site's roots to deeper, profoundly problematic dynamics at the heart of the endangered species protection regime. This chapter addresses how and why legal protection of endangered species is often contingent on enshrining, naturalizing, and even rendering invisible the destructive human approaches to environments and land use that imperil these species.

The continual expansion of unsustainable human-built environments has destroyed essential manatee habitat, particularly hot spring-fed marshlands that enable these warm-blooded mammals (Figure 14.1) to survive cold-water temperatures. As humans continue to deplete the Florida Everglades and other regional wetlands, the manatees have turned to alternative sources of winter warmth – hot-water effluent channels from several hydrocarbon-fueled power plants, including Big Bend. Though an endangered species finding sanctuary in the effluent of major greenhouse gas generators may sound counterintuitive, what led to the protection of these sites as state and federal wildlife preserves is equally ironic. Well-intentioned environmental activists, wielding one of the most powerful legal tools for protection of ecosystems – the Endangered Species Act – successfully pushed the US federal government to protect power plant effluent streams to safeguard manatees.

Contemporary work in environmental communication offers insights into a problematic, two-part construct that permeates Westernized and industrialized understandings of humanity's place within the "more-than-human world" (Abram, 1997). Under a dominant human/nature binary, humans are constituted as intrinsically, spatially, and legally different from – and superior to – "the natural." This case study illuminates how this binary, which underlies and animates dominant Western behaviors towards ecosystems and other species

Figure 14.1 An adult manatee and its calf. These large aquatic mammals, weighing up to 3,000 pounds, can live in both marine and freshwater systems.

Source: Photograph by Keith Ramos, from US Fish and Wildlife Service.

(Abram, 1997; Carbaugh, 1996; Coole & Frost, 2010; Milstein, 2016; Smith, 2008; Tsing, 2012), is enacted both legally and spatially in the case of endangered species. This chapter's analysis suggests that both cultural practices and legal regimes discursively separate the human sphere from the natural, while making places such as the Big Bend plant effectively "invisible" where this discursive separation fails. These *invisible ecocultural spaces* prop up a meaning system of human existence premised upon, and obscured by, unsustainable binary constructions (e.g., culture/nature, human/other animals, animal/habitat). However, human activity impacts the current state of the planet – from the ozone to the deepest reaches of the oceans. Likewise, our lives – including our cities, homes, and very bodies – are constituted by, and deeply entangled with, multitudes of forms of non-human life (e.g., as globally experienced now with COVID-19).

Environmental communication scholars illustrate ways this problematic human/nature binary is reproduced as common sense in institutional and everyday discourse (Carbaugh, 1996; Milstein, 2009; Milstein, Thomas, & Hoffmann, 2019). This literature illuminates a dominant globalizing anthropocentric approach to existence by which human mastery and the valuation of nature based on its extractive short-term utility to our species are prevailing practical and philosophical underpinnings to human–environment interactions. Complementary work in legal geography – a transdisciplinary research field investigating the spatially uneven workings of law – resonates with this research, often rooting its analysis in the concept of "property." For example, Macpherson (1962) argued that Locke's *Second Treatise* of 1689 (2011) marks the crystallization of the Western division of the human from the natural through logics of property. Specifically, Locke bases human society on a divinely granted right to accumulate property, the product of applying individual labor to a "nature" otherwise culturally constructed as inert and valueless (Macpherson, 1962).

Subsequent legal geographic work explores multifaceted ways both property law and cultural discourses about property help structure the ostensibly separate spheres of the human and the natural. Much of Blomley's work, for example, concerns ways the law constructs an understanding of "property as a bounded and territorialized thing, rather than a set of interlocking local obligations and relations" (2007, p. 2). By constructing portions of the planet as legally, physically, and politically separate realms, subject solely to the individual agency of property owners – and isolated from the influence of broader social and ecological networks – property has become an essential tool in the daily work of enforcing the human/nature binary.

Similarly, Graham (2011) argues that understanding the human place in the world through the lens of property is at the heart of current environmental crises. Western property law enables a "particular mindset, a paradigm that detaches people from place, viewing the world not as a network of relationships, but as two separate spheres: culture and nature" (p. 1). Graham argues, "Living in a world of modern property relations involves an acceptance that the real is unreal, that places are spaces, and that networks of complex relationships are commodifiable environmental products" (p. 11).

Statement of the problem

While the legal concept of property and dominant human/nature binary help clarify how anthropogenic ecological destruction has become normalized, neither sufficiently explains anomalous sites, such as the Big Bend endangered manatee sanctuary. Nor does it explain the broader, systematic tendencies within Western law to attempt (and largely fail) to protect specific endangered species such as the manatee while leaving unchallenged the dominant systemic human environmental practices that threaten all species. This chapter argues the law not only codifies the human and the natural as separate and distinct spatial categories, but in some cases also makes invisible spaces that otherwise could be clearly seen as neither purely human nor purely natural to protect that unsustainable binary and related vested interests and unsustainable lifeways (see, e.g., Carr & Milstein, 2018).

Findings

Unraveling the relationship between manatees and fossil-fuel power plants

Florida manatees (*Trichechus manatus latirostris*) are slow-moving, gentle relatives of the elephant that thrive in temperate shallow waters. As one of the original 78 species listed as facing extinction in the United States under the 1966 federal Endangered Species Preservation Act, Florida manatees face multiple individual and systematic anthropogenic threats. Rampant growth in unsustainable human-built environments along Florida's 1,350 miles of coastline has drained groundwater and drastically transformed a state that was 60% wetlands when European colonist settlers arrived. Manatees lack the blubber needed to inhabit cold waters, and destruction of their habitat has cut off their access to much of the warm water essential to their winter survival. In the unprecedented cold winter of 2010, 700 manatees were found dead, with at least 246 of the deaths directly attributed to cold stress, and the remaining likely attributable to the same cause. That winter, three power plants along Florida's east coast had performed scheduled shutdowns at the same time, halting their production of the warm-water effluent that manatees are now forced to rely on to survive.

In addition to this systemic threat of human destruction of habitat, boaters running watercraft into manatees pose an additional risk. Between 1958 and 2019, the number of boats registered in Florida grew from 100,000 to more than 960,000 (Florida Fish and Wildlife Conservation Commission, 2012; Florida Highway Safety and Motor Vehicles, 2019). As early as the 1940s, people drove boats into manatees so frequently that a propeller scar catalog was created as a system to identify individual manatees. Florida has since instituted speed limits in areas frequented by manatees, though these limits are not typically enforced (Rizzardi, 2011). Additional anthropogenic risks include harassment, including feeding, chasing, separating mother and calf, physically injuring them, and riding them. Another risk is red tide, naturally occurring toxic algae blooms exacerbated in frequency and severity due to anthropogenic climate disruption (Erdner et al., 2008; Wines, 2013; Zingone & Enevoldsen, 2000).

Because unsustainable human accretion has drastically diminished access to their spring-fed winter habitat, more than 60% of the Florida manatee population now relies on warm-water outfalls from power plants to survive, including those from the Big Bend Power Station near Tampa, owned and operated by a subsidiary of TECO Energy. The plant provides electricity to west central Florida via four coal-fired units and one hydrocarbon gas–fired unit.

Big Bend is part of a larger network of power-plant effluent channels now essential to manatee survival and incorporated into state and federal endangered species protection. Manatee dependence on power plants to survive became evident as early as 1977, when a partial shutdown of the Brevard County power plant was responsible for 38 cold stress deaths. In 1979, Florida adopted its first state-designated manatee protection zones, which came into effect during winter months and focused on areas where manatees gather in great numbers, including warm natural springs, power plant effluent, and other industrial sources (Florida Fish and Wildlife Conservation Commission, 2012). Shortly thereafter, power companies began to promote themselves as manatee protectors. In 1986, TECO opened a manatee viewing boardwalk (Figures 14.2 and 14.3) to the public at Big Bend. And, in 1993, TECO's Big Bend environmental education building and Manatee Viewing Center opened.

A 1997 proposal for deregulation or reorganization of Florida's power industry, which would have resulted in closure or intermittent operation of power plants critical for manatee winter survival, highlighted the need to address manatee dependence. In 1999, the three agencies charged with protecting manatee habitat under state and federal law – the Florida Fish and Wildlife Conservation Commission, the US Fish and Wildlife Service (USFWS), and the Florida Department of Environmental Protection – approved the first Power Plant Manatee Protection Plans. In 2000, the USFWS formed the Manatee Habitat Warm Water Task Force to address issues related to industrial facilities providing "habitat" to manatees during the cold season.

Figure 14.2 Entrance to the manatee viewing boardwalk, with power plant structures in the background.

Source: Photograph by John Carr and published with permission.

Figure 14.3 Boardwalk signs prohibiting individual human interference with manatees against the backdrop of the fossil fuel-burning Big Bend power plant.

Source: Photograph by John Carr and published with permission.

Legally contextualizing the fossil-fuel power plant manatee sanctuary

As anomalous as an endangered species sanctuary enabled by a fossil-fuel power plant may seem at first glance, creation of the sanctuary was the product of dominant legal approaches to endangered species protection at state and federal levels. These approaches treat certain human actions as capable of legal regulation – typically those involving individual human beings in "natural realms." In contrast, the expansion of property rights over critical habitat and ever-increasing development is typically treated as beyond the law's scope.

Two federal laws, the Marine Mammal Protection Act (MMPA) of 1972 and the Endangered Species Act (ESA) of 1973, protect Florida manatees. The Florida Manatee Sanctuary Act (FMSA) of 1978 also provides protection. The protection promised by these three laws is similar, reflecting the tendency of state environmental laws to mirror federal law. Specifically, the MMPA makes it illegal to "take" marine mammals without a permit. This means people may not "harass, feed, hunt, capture, collect, or kill any marine mammal or part of a marine mammal" (16 U.S.C. 1362 (13)). In turn, the MMPA defines harassment as:

> any act of pursuit, torment, or annoyance that has the *potential to injure* a marine mammal or marine mammal stock in the wild. . .; or that has the *potential to disturb* a marine mammal or marine mammal stock in the wild by causing disruption of behavioral patterns, including, but not limited to, migration, breathing, nursing, breeding, feeding, or sheltering.
>
> (16 U.S.C. 1362 (18)A) (emphasis added)

This provision could be interpreted as prohibiting any human activity that disrupts marine mammals' behavioral patterns. It could even extend to structural human disruption of marine mammals' ecosystems to the extent such disruptions impact their behavioral patterns and health.

Similarly, the ESA provides the potential for systematic protection of the ecosystems on which endangered species rely. Like the MMPA, it prohibits "taking" an endangered species, which section 3(19) defines as "to harass, harm, pursue, hunt, shoot, wound, kill, trap, capture, or collect, or to attempt to engage in any such conduct." Recognizing the importance of functioning ecosystems to the survival of endangered species, USFWS regulations interpret "taking" to include "habitat modification or degradation where it actually kills or injures wildlife by significantly impairing essential behavior patterns, including breeding, feeding, or sheltering" (50 CFR § 17.3).

Because these provisions potentially prohibit any human impairment of environments that endangered species rely on for survival, scholars such as Benson (2012) have argued the ESA could be "the strongest federal statute in the United States against species loss," as well as one of the most controversial (also see Scott et al., 2005; Freyfogle & Goble, 2009). In practice, the ESA's systematic protection of habitat has actually been very limited. At a statutory level, its protections are restricted to "critical habitat," namely, "specific areas within the geographical area occupied by the species . . . on which are found those physical or biological features (I) essential to the conservation of the species and (II) which may require special management considerations or protection" (16 USC § 1532(5)(a)(i)). Critical habitat, however, "shall not include the entire geographical area which can be occupied by the threatened or endangered species" (16 USC § 1532(5)(c)). Moreover, any potentially critical habitat may be excluded from statutory protection if there is an administrative determination that "the benefits of such exclusion outweigh the benefits of specifying such area as part of the critical habitat, unless . . . the failure to designate such area as critical habitat will result in the extinction of the species concerned" (16 USC § 1533(b)(2)).

The statute, thus, provides a minimalist approach to protecting habitat by limiting its reach to sites whose loss can be proven to lead to the imminent extinction of a species. It is frequently impossible, however, to conclusively argue that failure to list a given area will lead to extinction so long as there are other available areas of occupation. Thus this statutory approach has been described as "remarkably ignorant of ecological reality" in its frequent failure to actually protect habitat of sufficient quality and quantity to prevent extinction (Ruhl, 1990, p. 1401). Moreover, listing endangered species under the ESA and MMPA and using those statutes to protect habitat often depend on broader political, economic, and social agendas, including biases towards certain species and economic push-back from property owners (Agardy, Di Sciara, & Christie, 2011; Brown & Shogren, 1998; Klein et al., 2008; Possingham et al., 2002; Ruhl, 1999; Sheldon, 1997). Accordingly, the ESA creates a robust appearance of habitat protection while, in fact, minimizing the spatial extent of systematic protection of areas that endangered species rely on for survival.

Like the MMPA and ESA, Florida's Manatee Sanctuary Act provides the appearance of vigorous endangered species protection – largely following the framework of the federal statutes – while minimally regulating human activity that adversely impacts those species' habitat. Specifically, it largely limits its prohibitions to acts of individual human agency that could harm manatees while omitting systemic human-induced risks from its coverage. Thus, the Florida law focuses on restraining water vehicle speeds but does not prohibit habitat destruction, despite its stated "intent" to protect seagrass beds and other essential environments (Florida Administrative Code 68C-22.003).

In summary, these three statutes could have been written, interpreted, and applied to prohibit individual human activities that could harm individual manatees, as well as broader, more systematic human activities that negatively transform and structurally restrict ecosystems on which the species fundamentally relies. Unfortunately, political, social, and legal blowback against use of the ESA to protect habitat by curtailing expansion of human development has led federal regulators, and even activists, to take a much more cautious, minimalist approach to laws that on their face call for the holistic protection of ecosystems. Examples include such high-profile cases as the 1990s battle between

US Pacific Northwest loggers and environmentalists over protection of the northern spotted owl and efforts in the 1970s to bar construction of the Tellico Dam on the Little Tennessee River to protect the endangered snail darter fish (Flournoy, 1993; Yaffee, 1995). The resulting legal regime fails to protect species from extractive human encroachment and other destructive processes.

Discussion

The failure of endangered species protection regimes reflects deeper problematic spatial and legal discourses that affect more and more of the globe. As written and implemented, these statutes reflect the discourse of human/nature separation in such a way that this binary boundary line is both protected by and obscured within the language and logic of the law. While the Big Bend plant is an extreme example of the law's myopia, most human-modified environments are largely concealed from the limited purview of laws protecting habitat. The presence of these human-claimed and unsustainably engaged environments within the law is detectable only by the vast amount of habitat they remove from the law's grasp. At the same time, the law obscures such sites where the inseparability of the human and the "natural" is most unescapable. By treating these sites as neither "human" nor "natural," but rather as a third type of space – *spaces of invisibility* – the law conceals their systemic interrelationship, rendering them largely inaccessible to critique or legal redress.

The legal protection and naturalization of the Big Bend effluent stream is typical of this third space approach to habitat and ecological protection. The integration of its site into the legal regime of species protection resulted from a process initiated in 2000 when environmental activists and groups sued to force the USFWS and the US Army Corps of Engineers to comply with the habitat protection mandates of the ESA and MMPA. The court approved a settlement in 2001 (Wildlife Species Managed by the Fish and Wildlife Conservation Commission, n.d.). Subsequent enforcement litigation required the federal agencies to designate manatee refuges and sanctuaries throughout Florida. As a result, the USFWS identified 13 protection areas in eight counties, including the effluent streams of the Big Bend power plant and two other fossil-fuel-based power generation stations (the Bartow Electric Generating Plant near St. Petersburg and the Gannon Electric Generating Station near Tampa) (50 CFR § 17.108(c)). Shortly afterward, Florida designated the Big Bend site as a protected manatee sanctuary (Fla. Admin. Code 68C-22.013).

The resulting legal framework and its incorporation of power plant effluent streams is a powerful testament to how statutes like the ESA simultaneously promise robust and visible endangered species habitat protection while naturalizing the destruction of that very habitat. This process occurs through three primary dynamics. First, the human threat is framed as a function of individual recreational activity rather than of more fundamental systemic, unsustainable sociocultural choices about lifestyles, land use, and high energy consumption. For example, the regulations enacting the Manatee Sanctuary Act prohibit all waterborne activities, such as swimming, diving, snorkeling, and fishing, within manatee sanctuary boundaries (Fla. Admin. Code 68C-22.002(11)). In contrast, property owners' legal rights – and what they do to that property – are largely left intact with an explicit license to access and maintain property by water within designated sanctuaries so long as watercraft operate at idle speed. Likewise, federal regulations creating manatee protection areas focus solely on regulation of "waterborne activities" (50 CFR § 17.102). While these restrictions are important, given the significant number of manatees injured or killed by watercraft collisions, the law's almost exclusive focus on individual recreational activities obscures broader anthropogenic ecological-scale systemic threats.

Second, the regulatory framework fails to address structural threats to manatees and their habitat. While the regulations implementing Florida's Marine Sanctuary law prohibit new construction within designated sanctuaries (Fla. Admin. Code 68C-22.003), neither the environmentalist litigants nor federal and state regulators sought to use the broad language of the ESA or MMPA to address broader structural threats to manatee populations, namely, modern human development of wetlands. Rather, by incorporating power plants as integral parts of the manatee regulatory regime, the law tacitly and paradoxically naturalizes the elimination of wetland habitat as an invisible cost of expanding human settlements enabled by the same power plants.

This leads to the third problematic dynamic of the regulatory regime: At a time of accelerating climate crisis, the Big Bend sanctuary effectively enshrines an environmentally problematic type of electricity-generating power plant – one powered by carbon-based fossil fuels. The law treats Big Bend not only as unproblematic but also as an essential component of the ecosystem needed for endangered species survival, when in fact fossil-fuel-driven climate disruption has led to "red tides" lethal to manatees (Erdner et al., 2008; Wines, 2013; Zingone & Enevoldsen, 2000). Indeed, activists and regulators publicly express concerns about how contemporary interpretations of the statutes effectively lock fossil-fuel-fired power plants into the ecosystem, even amid environmental and climate protection movements to retire them. As one US Geological Survey biologist noted, "I'm pretty optimistic about the future of the manatees, except when I think about this monster we've created with their dependence on artificial warm water" (Lovgren, 2006, p. 2).

The practical difficulties of extracting carbon-based power plants from manatees' habitats without substantial harm is further complicated by the inherent conflicts among value systems of stakeholders – such as property owners and environmental activists – and uncertainty about how to create alternative viable habitat in the face of property ownership (Metcalf & Widener, 2011). That said, sites such as Big Bend bring into focus the pressing need to challenge the status quo that assumes ever-expanding extractive and destructive forms of development must be taken as an unproblematic baseline, while other non-human species must give way. Indeed, it is essential to envision and enact a more robust approach to species protections that treats neither the human/

nature binary nor property as unchangeable, and instead leads societies to reclamation and restoration of essential and endangered habitat.

Conclusion

This case illustrates multiple ways contemporary environmental laws blind societies to systemic-scale anthropogenic destruction by institutionalizing and enforcing an ecoculturally dominant human/nature binary. At the same time, the case illustrates ways property law serves as both background to and horizon for ecosystem protection by maintaining the border between human and natural realms in the division between public and private space. For instance, the Big Bend sanctuary extends only as far as the publicly held wetlands and effluent channel lying between the power plant and the viewing center, but not within the actual power plant site. This prevents the reach of both federal and state endangered species regulations across corporate private property lines and into destructive corporate practices (50 CFR § 17.108(c), Fla. Admin. Code 68C-22.013). Similarly, by articulating species protection as a matter of limiting individual agency – whether limiting boat speed or barring contact with individual manatees – statutes reproduce a vision of humanity as only nominally engaged with ecosystems and only within the scope of recreational or touristic activities. In doing so, the law reduces its vision of interaction between the human and the natural to places and times when individuals leave a supposedly purely human sphere to enter, and subsequently threaten, a more-than-human realm. And in this case, the law obscures both the fossil-fuel-burning power plant and the human development and resulting environmental destruction it enables.

This case also offers a powerful example of how the law creates zones of invisibility in places where the inherent inseparability of the "human" and "natural" spheres is most spectacularly revealed (Carr & Milstein, 2018). Graham argues, "What we see in a landscape is not a detached and separated physical realm, 'the land,' but ourselves, our practices and our law – a lawscape" (2011, p. 5). Sites such as Big Bend do not fit neatly within the lawscapes of either nature or culture. Accordingly, the law creates a third type of lawscape, one that actively obscures places where the inseparability of humans and ecology is particularly visible.

This regulatory regime to protect manatees, their habitat, and their ecosystems also has been used to legally, and thus culturally, make invisible the greatest threats to all three. By failing to regulate the expansion of unsustainable development into wetlands and other essential habitat, the law juridically obscures those systemic ecologically destructive actions and processes. In turn, this legal invisibility allows anthropogenic habitat annihilation to continue as an unquestioned and immutable manifestation of the logics of nature as property.

The legal creation of such a paradoxical zone of invisibility is neither coincidental nor unique. Rather, it obscures and perpetuates the fundamental cultural and existential contradictions at the heart of the dominant human/nature binary. The zone of invisibility is essential precisely because the underlying binary is itself illusory and inherently unstable. In fact, every human body, activity, and place is profoundly informed by physical geographies, biosystems, climates, and micro-climates, as well as by non-human life-forms, including bacteria, toxins, parasites, and viruses. Likewise, every non-human life and "wild" place is integrally impacted by human influence, which is accelerating during the current epoch of profit-as-progress and unrestrained exploitation (Moore, 2015). Human existence is premised on and inseparable from the ecological webs of natural existence.

In contrast, a world defined by property and its underlying human/nature binary premise demands "an acceptance that the real is unreal" (Graham, 2011, p. 11). Today, dominant discourses define humanity in terms of a narrow range of high resource-intensive activities premised on the construction of humanity's difference from the biodiversity of which humans are one part. These activities range from treating portions of the Earth as private property to defining personal identity in terms of patterns of conspicuous overconsumption and structuring daily life and social networks around high-waste activities that drive species extinctions and ecological obliterations (see Milstein & Castro-Sotomayor, 2020). Indeed, more than 40 years ago, Basalla (1979, p. 39) recognized that humanity and civilization have become equated with the consumption of energy, with dominant discourses warning that any reduction in consumption will signal a return to barbarism.

Thus, the fundamental meaning of humanness is linked to destructive extractive practices that maintain nature's "otherness." This is revealed through built environments premised on disregarding the presence and/or essential role of other processes and forms of life. Typically, these ever-expanding unsustainable human accretions disturb and destroy ecosystems, are guarded with poisons, and sealed off with asphalt and concrete. Within them, non-human life is to be incorporated only as mastered, bounded, and controlled elements in a human-dominated lifeway. When the human/nature binary is shaken by reality, however, cracks quickly appear.

Because the hierarchical separation of the "human" and "natural" is untenable, these categories constantly collapse into one another. This collapse can be seen in non-human life-forms entering the human world without humanity's permission (e.g., the reentry of many non-human species into our cities during COVID-19 lockdowns (Singh, 2020)) and *vice versa* (e.g., a pandemic-level virus emerging from markets that exploit wildlife or climate-disrupted, drought-hungry bears entering cities). It can be seen in industrial practices changing the composition of the atmosphere, lands, and waters, as well as in the catastrophic impact of profit-driven climate crises on fires, floods, and temperature levels. Indeed, the stubborn refusal in much of the industrialized world to wean from dependence on carbon-based energy suggests it may be easier to envision and accept the willful extinction (or extermination) of humans and other species than to reimagine ways of living that are premised on the knowledge that humanity is just one interrelated part of our shared biosphere.

A host of strategies, tactics, denials, and discourses are increasingly mobilized to mask this intensifying ecocultural crisis. This chapter suggests that one such strategy is the creation of legally invisible "third spaces." These third spaces are zones in which the law treats the very things it is intended to regulate as outside its purview, including systemic anthropogenic ecological destruction. The law does so through selective and limited focus and omission. This takes several forms in the case of Big Bend. Most obviously, by interpreting the ESA and other relevant statutes as simultaneously protecting endangered species while remaining silent about systems-level causes of their endangerment, the law obscures systemic causal human threats to species survival. Likewise, by positing fossil-fuel-fired power plants as part of a network of endangered species sanctuaries, the regulatory regime establishes such sites as preeminent third spaces of invisibility. It obscures their destructive environmental impacts and reduces their legally "visible" components to a limited number of beneficial warm-water outputs offering highly limited refuge for endangered wildlife.

Of most concern is how these third spaces of invisibility function and are upheld within the same mechanisms established for ecosystem and species protection. As such, they render invisible, and thus unproblematic, dominant societal practices that imperil the biosphere. For the manatees, these concerns have intensified. In 2017, the USFWS downlisted Florida manatees from "endangered" to "threatened," largely in response to pressure from corporate lobbyists and the Pacific Legal Foundation, a free-market legal advocacy group representing property owners on Florida's Gulf Coast (Bryan & Fleshler, 2016, Frazier, 2020). The erosion of already minimal systems of protection, combined with the continuing accretion of unsustainable forms of human development fueled by fossil-fuel-burning power plants, highlights the precarious existence of species and their ecosystems within legal and property regimes that protect and make the human/nature binary concrete.

Acknowledgments

This chapter draws on research previously published in Carr, J., and Milstein, T. (2018). Keep burning coal or the manatee gets it: Rendering the carbon economy invisible through endangered species protection. *Antipode*, *50*(1), 82–100. It is published with permission from John Wiley & Sons, Inc.

References

Abram, D. (1997). *The spell of the sensuous: Perception and language in a more-than-human world* (1st ed.). London: Vintage Books.

Agardy, T., Di Sciara, G. N., & Christie, P. (2011). Mind the gap: Addressing the shortcomings of marine protected areas through large scale marine spatial planning. *Marine Policy*, *35*(2), 226–232.

Basalla, G. (1979). Energy and civilization. In *Science, technology and the human prospect* (Vol. 4, pp. 39–52). Oxford, UK: Pergamon Press.

Benson, M. H. (2012). Intelligent tinkering: The Endangered Species Act and resilience. *Ecology and Society*, *17*(4), 1–8.

Blomley, N. (2007). Making private property: Enclosure, common right and the work of hedges. *Rural History*, *18*, 1.

Brown, G. M., & Shogren, J. F. (1998). Economics of the Endangered Species Act. *Journal of Economic Perspectives*, *12*(3), 3–20.

Bryan, S., & Fleshler, D. (2016). Manatee may lose endangered status. Retrieved from www.sun-sentinel.com/local/broward/fl-manatees-reclassify-petition-20160107-story.html

Carbaugh, D. (1996). Naturalizing communication and culture. In J. G. Cantrill & C. L. Oravec (Eds.), *The symbolic earth: Discourse and our creation of the environment* (pp. 38–57). Lexington, KY: University Press of Kentucky.

Carr, J., & Milstein, T. (2018). Keep burning coal or the manatee gets it: Rendering the carbon economy invisible through endangered species protection. *Antipode*, *50*(1), 82–100.

Coole, D., & Frost, S. (Eds.). (2010). *New materialisms: Ontology, agency, and politics*. Chapel Hill, NC: Duke University Press.

Endangered and Threatened Wildlife and Plants, 50 CFR 17 (2001).

Endangered Species Act of 1973, 16 USC § 1532–1533 (1973).

Erdner, D. L., Dyble, J., Parsons, M. L., Stevens, R. C., Hubbard, K. A., Wrabel, M. L., . . . Trainer, V. L. (2008). Centers for oceans and human health: A unified approach to the challenge of harmful algal blooms. *Environmental Health*, 7(2), S2.

Florida Administrative Code, 68C-22.013 (2005).

Florida Fish and Wildlife Conservation Commission. (2012). Manatee timeline in Florida. Retrieved from http://myfwc.com/education/wildlife/manatee/timeline

Florida Highway Safety and Motor Vehicles. (2019). 2019 alphabetical vessel statistics by county. Retrieved from www.flhsmv.gov/motor-vehicles-tags-titles/vessels/vessel-owner-statistics/

Flournoy, A. C. (1993). Beyond the spotted owl problem: Learning from the old-growth controversy. *Harvard Environmental Law Review*, *17*, 261.

Frazier, M. (2020, April 2). Nobody knows how to wean manatees off coal plants. *Bloomberg Businessweek*. Retrieved from https://www.bloomberg.com/news/features/2020-04-02/nobody-knows-how-to-wean-manatees-off-coal-plants

Freyfogle, E., & Goble, D. D. (2009). *Wildlife law: A primer*. Washington, DC: Island Press.

Graham, N. (2011). *Lawscape: Property, environment, law*. London: Routledge.

Klein, C. J., Chan, A., Kircher, L., Cundiff, A. J., Gardner, N., Hrovat, Y., . . . Airame, S. (2008). Striking a balance between biodiversity conservation and socioeconomic viability in the design of marine protected areas. *Conservation Biology*, *22*(3), 691–700.

Locke, J. (2011). *Second treatise of government*. Seaside, OR: Watchmaker Publishing.

Lovgren, S. (2006, October 20). Manatees seek power plants, warm springs as safe havens. *National Geographic News*. Retrieved from http://news.nationalgeographic.com/news/2006/10/061020-manatees.html

MacPherson, C. B. (1962). *The political theory of possessive individualism: Hobbes to Locke*. Oxford, UK: Oxford University Press.

Marine Mammal Protection Act of 1972, 16 U.S.C § 1362 (1972).

Metcalf, S. S., & Widener, M. J. (2011). Growing Buffalo's capacity for local food: A systems framework for sustainable agriculture. *Applied Geography*, *31*(4), 1242–1251.

Milstein, T. (2009). "Somethin' tells me it's all happening at the zoo": Discourse, power, and conservationism. *Environmental Communication: A Journal of Nature and Culture*, *3*(1), 24–48.

Milstein, T. (2016). The performer metaphor: "Mother Nature never gives us the same show twice." *Environmental Communication, 10*(2), 227–248.

Milstein, T., & Castro-Sotomayor, J. (Eds.). (2020). *Routledge handbook of ecocultural identity*. London: Routledge.

Milstein, T., Thomas, M., & Hoffmann, J. (2019). Dams and flows: Immersing in Western meaning systems in search of ecocultural reflexivity. *Environmental Communication, 13*(1), 104–117.

Moore, J. W. (2015). *Capitalism in the web of life: Ecology and the accumulation of capital*. London: Verso Books.

Possingham, H. P., Andelman, S. J., Burgman, M. A., Medellín, R. A., Master, L. L., & Keith, D. A. (2002). Limits to the use of threatened species lists. *Trends in Ecology & Evolution, 17*(11), 503–507.

Rizzardi, K. W. (2011). Toothless? The endangered manatee and Florida Manatee Sanctuary Act. *Research Gate, 24*. Retrieved from www.researchgate.net/publication/256011263_Toothless_The_Endangered_Manatee_and_Florida_Manatee_Sanctuary_Act

Ruhl, J. B. (1990). Regional habitat conservation planning under the Endangered Species Act: Pushing the legal and practical limits of species protection. *Southwestern Law Journal, 44*, 1393.

Ruhl, J. B. (1999). How to kill endangered species, legally: The nuts and bolts of Endangered Species Act "HCP" permits for real estate development. SSRN Scholarly Paper ID 177928. Social Science Research Network. Retrieved from http://papers.ssrn.com/abstract=177928

Scott, J. M., Goble, D. D., Wiens, J. A., Wilcove, D. S., Bean, M., & Male, T. (2005). Recovery of imperiled species under the Endangered Species Act: The need for a new approach. *Frontiers in Ecology & the Environment, 3*, 383–389.

Sheldon, K. P. (1997). Habitat conservation planning: Addressing the Achilles heel of the Endangered Species Act. *New York University Environmental Law Journal, 6*, 279.

Singh, M. (2020, April 22). Emboldened wild animals venture into locked-down cities worldwide. *The Guardian*. Retrieved from www.theguardian.com/world/2020/mar/22/animals-cities-coronavirus-lockdowns-deer-raccoons

Smith, N. (2008). *Uneven development: Nature, capital, and the production of space* (3rd ed.). Athens, GA: University of Georgia Press.

Tsing, A. (2012). Unruly edges: Mushrooms as companion species for Donna Haraway. *Environmental Humanities, 1*(1), 141–154.

Wildlife Species Managed by the Fish and Wildlife Conservation Commission. (n.d.). Retrieved from http://myfwc.com/wildlifehabitats/managed/manatee/

Wines, M. (2013, April 6). Algae bloom in Florida kills record number of manatees. *New York Times*.

Yaffee, S. L. (1995). Lessons about leadership from the history of the spotted owl controversy. *Natural Resources Journal, 35*, 381.

Zingone, A., & Enevoldsen, H. O. (2000). The diversity of harmful algal blooms: A challenge for science and management. *Ocean & Coastal Management, 43*(8), 725–748.

15 Conservation on private land

The Endangered Species Act

Joe Kerkvliet, Christian Langpap and Jason Shogren

The comprehensive environmental laws passed by Congress in the 1970s are command-and-control regulations by which "authorities simply mandate the behavior in law, then use whatever enforcement machinery – courts, police, fines – to get people to obey the law" (Field & Field, 2013, p. 206), as opposed to financial and other types of incentives. State and local authorities mostly took similar "police power" approaches to environmental policy, including comprehensive and master planning, zoning, development review, and permitting (Wright & Czerniak, 2000).

The 1973 Endangered Species Act (ESA), designed to protect and recover imperiled species and the ecosystems they depend on, relied predominately on command-and-control. State and local governments, concerned about biodiversity loss, the disappearance of ecologically important land and open space, followed suit with their own versions of command-and-control (Wright & Czerniak, 2000).

Command-and-control regulations have advanced environmental objectives, but not without raising conflicting issues. Such regulations can lead to more expensive improvements than necessary for the proposed environmental outcomes (Field & Field, 2013). There can be unintended consequences if private agents responding to command-and-control behave in ways that produce results that conflict with the intent of the law. In some cases, existing regulations may actually have slowed progress towards achieving environmental goals and caused voters to voice opposition to regulations and/or taxes (Lueck & Michael, 2003; Shaffer et al., 2006). Economic and political opposition has been strong and often threatens the strength, and even the existence, of environmental laws (Scott, Gogle, & Davis, 2006). Charges of unconstitutional takings of private property are rife (Salzman & Thompson, 2017). There have been "train wreck" confrontations involving landowners, property rights advocates, environmental activists, years of litigation, and even federal marshals (Young, 2001; Alexander, 2018). And there are perceptions among some voters and policy makers that environmental regulations are ineffective in protecting important lands and threatened and endangered species (Wright & Czerniak, 2000, Scott et al., 2006; Shaffer et al., 2006).

To buffer these conflicting issues, policy makers and nongovernment organizations have pursued new programs to protect and restore endangered species and the habitats on which they depend by aligning biodiversity protection with the interests of private agents, especially landowners. The agencies directly administering the statute, the US Fish and Wildlife Service (USFWS) and the National Marine Fisheries Service (NMFS), have developed and promoted Habitat Conservation Plans, Safe Harbor Agreements, and Candidate Conservation Agreements with Assurances. Governments and more than 1,700 national, regional, and local land trusts have actively purchased or negotiated conservation easements on more than 40 million acres of private land, often to protect and restore threatened or endangered species (Langpap, Kerkvliet, & Shogren, 2018). Land trusts are nongovernmental organizations whose efforts are strongly supported by tax incentives and government grants (Parker, 2008).

Statement of the problem

When a species is listed as endangered or threatened under the ESA, restrictions begin on uses of both public and private land. Some restrictions differ between public and private land, in law or in application. Still, of the over 1,600 threatened and endangered species in the United States, fewer than half are found on federal land. About two-thirds can be found on privately owned land (Groves et al., 2000; Evans et al., 2016). Private land, especially in Western states, is disproportionately important relative to its size because it often contains the best access to water and the most fertile soils (Bean, Bonnie, Male, & Searchinger, 2003).

There is substantial evidence that ESA regulations or the threat of regulations have induced some landowners to pre-emptively modify habitat to discourage imperiled species from occupying it (Lueck & Michael, 2003; Langpap, Kerkvliet, & Shogren, 2018). Since private owners are almost always able to enhance and conserve – or degrade and destroy – the biological resources on their land, it is important that they be considered in designing policies to reduce extinctions and enhance recovery of threatened and endangered species.

The Endangered Species Act and private landowners

Many actions and restrictions are mandated when a species is listed (Matsumoto, Pike, Turner, & Wah, 2003) and may be enforced by the USFWS and NMFS. This chapter focuses on regulations affecting land use and land management. Section 4 of the ESA mandates determination of critical habitat. Section 6 outlines cooperation with and grant opportunities for states, tribes, and landowners. Section 9 prohibits the "take" of listed species, and Section 10, enacted in 1982, recognizes exceptions to "take" prohibitions and authorizes "incidental take permits" if landowners develop an approved Habitat Conservation Plan.

Landowners and property rights advocates perceive Section 9's "take" prohibition as interfering with management prerogatives (Salzman &Thompson, 2017). At the extreme, they claim that Section 9 violates their Fifth Amendment protections against uncompensated taking of private property for public purposes. The ESA defines "take" as "to harass, harm, pursue, hunt, shoot, wound, kill, trap, capture, or collect" threatened and endangered species. Later, the agencies administratively defined "harm" to include impairing habitat in ways that might disrupt feeding, breeding, or other essential behavior patterns (Salzman & Thompson, 2017). In a 1995 case involving restrictions on logging private land containing habitat for northern spotted owls (*Babbitt v. Sweet Home Chapter of Communities for a Great Oregon*), the US Supreme Court upheld this definition (see Salzman &Thompson, 2017 for details).

Political opposition to the ESA and ESA innovations

Although the ESA passed Congress with little opposition in 1973, it soon encountered widespread, strong, and persistent criticism. While some reactions stemmed from the statute's record of curtailing federal projects, private landowners and property rights advocates zeroed in on its potential threats to their management options and Fifth Amendment protections. The need for statutory changes was motivated by strong political opposition that threatened to weaken the law in the process of its 1994 congressional review (Ruhl, 2004; Scott et al., 2006). In addition, there was evidence that restrictions on land use sometimes produced perverse results, such as inducing landowners to pre-emptively modify their property to discourage such species (Brown & Shogren, 1998). For example, the ESA-listed red-cockaded woodpecker (*Dryobates borealis*) requires large tracts of older southern pine for breeding and nesting. Lueck and Michael (2003) found that Southern timberland owners clear-cut more trees at younger ages when their land was more likely to be occupied by the woodpeckers.

By 1994, mounting political opposition had raised the prospect of congressional weakening of the statute. Interior Secretary Bruce Babbitt proposed refocusing the agencies' efforts and administrative changes, instead of modifying the ESA (O'Keefe, 1994; Ruhl, 2004; Scott et al., 2006). The eventual changes included increased emphasis on, and redesign of, programs in ways that paid attention to landowner concerns and motivated them to engage in conservation efforts, such as Habitat Conservation Plans, Safe Harbor Agreements, and Candidate Conservation Agreements with Assurances. All these programs contain three common elements: a conservation standard, some flexibility in that the agency and landowners negotiate each document, and some regulatory certainty for landowners.

Safe Harbor Agreements

Safe Harbor Agreements are designed to encourage collaboration with conservation-minded landowners. They are voluntary contracts between a non-federal property owner and the USFWS or the NMSF. Under such an agreement, the landowner consents to actions contributing to the recovery of listed species. These actions include reductions in habitat fragmentation, habitat restoration, increases in species population, buffers for protected areas, and testing new conservation strategies (USFWS & NMFS, 1999b).

Landowners receive assurances that the agency will require no additional actions other than those in the agreement and that the property may be returned to its original, baseline condition at the end of the contract. These incentives are affected by the agency's Enhancement of Survival Permits, which essentially exempt the landowners from Section 9's "take" prohibitions. To be approved, the conservation standard a Safe Harbor Agreement must provide is "net conservation benefits" to the covered species.

Final rules for Safe Harbor Agreements were published in 1999. By 2005, the USFWS reported 32 agreements covering 3.6 million acres and 36 threatened and endangered species (USFWS, 2005). In 2017, Bean (2017) reported at least 100 agreements.

The Oregon chub story

The 2015 delisting of the Oregon chub (*Oregonichthys crameri*) after 23 years' effort is given as an example of how innovations in regulatory approaches to biodiversity protection can work (American Fisheries Society, 2015). The Oregon chub is a small floodplain minnow endemic to Oregon's Willamette River basin. Sixty percent of its habitat is on private land. When field surveys revealed it had disappeared from most of its historic locations, it was listed as endangered in 1993. The surveys identified the major causes of extinction risk as altered river flow regimes and introduced non-native fish.

Recovery efforts involved all three aspects of Babbitt's innovations in biodiversity protection. The Fish and Wildlife Service cooperated with the US Army Corps of Engineers to adjust water releases from the dams it manages on Willamette River tributaries to more closely mimic pre-dam temperature and flow regimes. Oregon Fish Wildlife and Parks recruited Willamette Valley landowners to allow re-introductions of the chub. The Safe Harbor Agreements provided assurances against future land use restrictions and the option to return to baseline conditions at the end of the 10- to 30-year agreements. Finally, the McKenzie River Trust purchased private property with existing chub populations, using funds from private donors and the Bonneville Power Administration.

In 2010, the USFWS downlisted the chub from endangered to threatened. In February 2015, it became the first fish ever delisted because of recovery. Prior to 2015, fish species had been delisted, but because of extinction, not recovery.

Candidate conservation agreements with assurances

Candidate Conservation Agreements with Assurances are another Babbitt-inspired innovation. Candidate species are those for which the agency has decided a listing is "warranted but precluded" by more urgent needs, or if there has been progress towards the species' recovery (Matsumoto et al., 2003). While the ESA's core protections are reserved for threatened and endangered species, these agreements provide a means for eliciting private landowners' efforts in pre-listing conservation efforts.

In 1999, the Fish and Wildlife Service had identified 154 candidate species and 69 proposed species (USFWS & NMFS, 1999a). In 2007, there were 280 candidate species (Paez, 2009). Extensive litigation, court orders, dedicated funding, and USFWS efforts have reduced the number of candidate species to 30 in 2016 and 37 in 2019 (USFWS, 2016, 2019).

There are several potential advantages to encouraging pre-listing conservation measures (Donlan, 2015; Li & Male, 2013). First, they may be less expensive and/or available before listing. For example, if a species occupies an area prior to listing, it will probably be less expensive to maintain its occupancy rather than to pay for re-introduction. Second, pre-listing measures may be more available than post-listing measures. For example, urban development practically destroys the possibility of any restoration. Third, if the need to list can be avoided, property owners can maintain land management flexibility. Fourth, if the need to list can be eliminated, the expense and controversy surrounding critical habitat designation can be avoided. Finally, landowners may be less inclined towards pre-emptive habitat modifications, even though such actions would not constitute illegal "take" for a candidate species.

In addition to these advantages for species conservation, Candidate Conservation Agreements with Assurances offer incentives to non-federal landowners. Specifically, those who voluntarily commit to implement conservation measures receive assurances that additional conservation measures will not be required and that additional land use restrictions will not be imposed if the species become listed in the future. Participants who fulfill the terms of their agreements receive an Enhancement of Survival Permit exempting them from the ESA's "take" provisions if the species later gets listed (USFWS & NMFS, 1999a).

By 2005, there were 10 Candidate Conservation Agreements with Assurances (USFWS, 2005). By 2013, Li and Male (2013) found 26, and two years later Nielson (2015) found 30. Several more have been issued since then, but a complete compilation has not been completed.

Research on the characteristics, implementation, and effectiveness of these agreements is in initial stages. Li and Male (2013) point out that some Candidate Conservation Agreements with Assurances, as written, are likely to provide benefits, but they also may authorize (through Enhancement of Survival Permits) activities that "take" the species. Li and Male (2013) found that a large majority of the agreements potentially provided for a mix of activities, some of which enhance species restoration and some of which may do the opposite. Only 12% of the 26 they examined provided unambiguous conservation benefits for the species.

Li and Male's (2013) work provides a useful starting point for research on the total effects of Safe Harbor Agreements and Candidate Conservation Agreements with Assurances. In promoting conservation efforts by private property owners, these programs provide incentives for landowners. Some incentives unambiguously promote species conservation; others may not. Still other incentives, such as those for oil and gas development, are likely to be harmful to the species but provide landowners with incentives to participate in other conservation initiatives.

Nielson (2015) reported 30 Candidate Conservation Agreements with Assurances in 19 states targeting 48 species, including seven candidate species. Those agreements covered at least 3.35 million acres. The largest agreement applies to 1.5 million acres in New Mexico with two targeted candidate species, the dunes sagebrush lizard (*Sceloporus arenicolus*) and the lesser prairie chicken (*Tympanuchus pallidicinctus*). The smallest is a one-acre Arizona agreement targeting the Page springsnail (*Pyrgulopsis morrisoni*). Neilson found that 70% of the agreements provided for construction or alteration of infrastructure; 67% provided for non-target species management, including predator control, weed control, and livestock management; 67% also provided for land use restrictions, including ones affecting farming practices, chemical use, hunting, and other recreational activities; and 40% contained provisions for reintroduction or population augmentation of the target species.

The fluvial Arctic grayling story

Until the 1930s, stream-dwelling Arctic grayling (*Thymallus arcticus*) occupied the entire Missouri River drainage above the Great Falls of the Missouri River. By the 1940s, the grayling's occupied habitat was shrinking due to dam-building, irrigation, non-native fish stockings, and declining water quality. In 1991, the Center for Biological Diversity petitioned the USFWS to list the fish, and in 1994, the agency gave it "warranted, but precluded" candidate species status. In 2003 and 2004, the center sued. However, the agency decided in 2007 that the grayling did not warrant listing. In the midst of this serial litigation, Montana Fish, Wildlife and Parks, USFWS, Montana Department of Natural Resources, and the US Department of Agriculture's Natural Resources Conservation Service announced a Candidate Conservation Agreement with Assurances for the upper half of the Big Hole watershed (MFWP, 2006). The agreement listed four broad goals: restoration of instream flows; riparian habitat restoration; elimination of fish entrainment (being trapped in irrigation infrastructure); and removal of migration barriers. By August 2006, the Big Hole agreement had issued "certificates of inclusion" to 33 landowners owning about half of the 300,000-acre watershed. It also included another 8,000 acres of state-owned land. By 2008, more than $4 million of public and private money had been spent on projects such as riparian fencing, willow plantings, and relocation of a Big Hole tributary (Stumpf, 2008). One $226,000 project diverted a Big Hole tributary, Rock Creek, from its constructed irrigation channel, which ended in a hayfield, to its original channel, which joined the Big Hole (Lamothe & Peterson, 2007).

In 2014, the USFWS determined the Arctic grayling did not warrant ESA protection. It justified the decision, in part, by reference to the conservation efforts of federal and state agencies, private landowners, and conservation groups (MFWP, 2018). However, in 2018 the US Ninth Circuit Court of Appeals partially reversed the agency's decision, holding that it had acted in "an arbitrary and capricious" manner. The ruling cited climate change factors and did not address the role of the Candidate Conservation Agreement with Assurances (US Court of Appeals, 2018).

In 2018, Montana Fish, Wildlife and Parks announced a draft Candidate Conservation Agreement with Assurances for Arctic grayling in Montana's Centennial Valley, an upper Missouri River drainage 60 miles southeast of the Big Hole (MFWP, 2018; US Fish and Wildlife Service, 2018). The document concerned Arctic grayling previously believed to be extirpated on non-federal land (MFWP, 2018, p. 1) and proposed essentially the same conservation measures as the Big Hole agreement.

The Big Hole River agreement bolstered the USFWS case to not list the grayling. but whether it has sufficiently aided the fish to remove threats to its existence is uncertain (Stumpf, 2008; Tennant, Magee, Cayer, & McCullough, 2011). One critic cited the Rock Creek project, saying that its channel restoration project was insufficient because climate change had reduced stream flows and often left the restored channel without water (Stumpf, 2008).

Are conservation efforts on private land effective?

To date, the evidence needed to answer the question of effectiveness is scant for both Safe Harbor Agreements and Candidate Conservation Agreements with Assurances. The Safe Harbor Agreement conservation standard is "net conservation benefits" for the targeted imperiled species, while the Candidate Conservation Agreements with Assurances aim is to obviate the need to list the target species (USFWS & NMFS, 1999a, 1999b).

Safe Harbor Agreements are credited by fisheries biologists as one of the drivers of the 2015 delisting of the Oregon chub (American Fisheries Society, 2015). Bean (2017) describes the success of several Safe Harbor Agreements in promoting red-cockaded woodpecker populations, but that species has not been delisted as of 2020. These instances provide some evidence that such agreements result in positive conservation outcomes. But they fall short of providing statistical evidence that they work.

Arguably, Candidate Conservation Agreement with Assurances contributed to avoiding a "train wreck" listing of the greater sage grouse (*Centrocercus urophasianus*) (Wilson, 2014), and the USFWS has used putative Big Hole Candidate Conservation Agreements with Assurances accomplishments to justify removing ESA listing protection for the fluvial Arctic grayling. Nielson (2015) found documentation that those conservation measures justified agency decisions to not list three species as candidates for listing. Again, there appears to be no statistical evidence of the effectiveness of these agreements in achieving ESA objectives, and this remains a topic for research.

One impediment to providing solid evidence is the difficulty of obtaining the prerequisite data on both types of agreements. Doremus (2006, p. 204) states "Even for simple bean counting we lack well-organized, accessible databases. . . [making] it extraordinarily difficult to undertake quantitative studies of ESA implementation." For Safe Harbor Agreements, Bean (2017) reports that the agency's conservation plans database is "incomplete, out of date, and sometimes wrong" (p. 9). Nielson's (2015) had similar concerns with Candidate Conservation Agreements with Assurances. Neilson was able to locate final or draft documents for what was thought to be all extant Candidate Conservation Agreements with Assurances, but the researcher had to contact landowners directly to obtain a handful of those documents. However, Nielson's efforts to obtain important documents pertaining to annual reports, monitoring, and landowner participants were often unsuccessful. Without this sort of information, Nielson sometimes could not determine even basic information, such as the total acreage enrolled in some agreements.

Discussion

These agreements are contracts among multiple parties with diverse, and often conflicting, objectives. As such, they must address tradeoffs on several margins, while the final documents must meet Safe Harbor Agreements or Candidate Conservation Agreements with Assurances conservation standards. Failure to address tradeoffs would risk agencies' disapproval and exposure to environmental organizations' lawsuits.

Foremost and legally, these agreements are designed to remove or reduce landowners' risk of further regulation if the target species becomes more imperiled or a new threat is recognized (Safe Harbor Agreements) or if it becomes listed (Candidate Conservation Agreements with Assurances). Evidence suggests that this incentive works in some, but not all, situations (Sorice et al., 2013; Langpap, 2004), and some analysts suggest that removal of the regulatory certainty provided by these agreement would result in a dramatic decrease in landowner participation (Doremus, 2006, p. 199).

Neither type of agreement offers direct monetary incentives. Research shows wide variation in the importance of monetary incentives to landowners, but such incentives nearly always matter somewhat (Sorice et al., 2013; Bastian, Keske, McLeod, & Hoag, 2017). Candidate Conservation Agreements with Assurances must identify funding for the projects proposed by the contract (USFWS, 2003), but Nielson (2015) found that is not always done. For some agreements, funding for specific projects seems readily available from public (e.g., the Agriculture Department's Natural Resources Conservation Service) and private sources. Some projects, such as fencing and more efficient irrigation infrastructure, might provide direct financial benefit to landowners.

Research strongly shows that landowners are loathe to cede decision-making over land management practices (e.g., Sorice et al., 2013; Fischer, 2004; Bastian et al., 2017). Sorice et al. (2013) found landowners prefer conservation programs that give them complete decision-making responsibilities and would require monetary compensation even to share these responsibilities as part of a conservation contract.

Conversely, conservation-oriented agents (e.g., USFWS and administrators of Candidate Conservation Agreements with Assurances) may hold strong views about the most effective means to accomplish conservation objectives. The tradeoff is that conservation-oriented agencies risk losing landowner participation if they embed these views too strongly in the contract (Sorice et al., 2013). Further, contract provisions that are too precise and restrictive in the "what, where, when, and how" of conservation practices risk losing the property-specific ecological knowledge that many landowners possess (Parker, 2004). Of course, if the contract yields all decision-making authority to landowners, there is a risk that conservation efforts will be severely reduced.

Conclusion

The purposes of the ESA are unlikely to be achieved without actively engaging the resources of private land and the support of private landowners. As enacted in 1973 and amended in 1982, the statute predominately applied command-and-control regulations to both public and private land. Experience teaches that more positive and innovative approaches towards private land are needed. The Safe Harbor Agreement and the Candidate Conservation Agreement with Assurances are two such approaches designed to attract landowners with flexible contract terms and less regulation. However, we have much to learn about the effectiveness of these agreements and how to promote the broad landowner participation needed to achieve the ESA's purpose of providing "a means whereby the ecosystems upon which threatened and endangered species depend may be conserved" (USFWS, n.d.).

Acknowledgments

This chapter draws on research previously published in Langpap, C., Kerkvliet, J., & Shogren, J. F. (2018). The economics of the US Endangered Species Act: A review of recent developments. *Review of Environmental Economics and Policy*, *12*(1), 69–91.

References

Alexander, K. (2018, July 20). The Klamath conflict. *San Francisco Chronicle*. Retrieved from www.sfchronicle.com/news/article/The-Klamath-conflict-Water-war-along-13089350.php

American Fisheries Society. (2015). Q and A: The success story of the Oregon chub: An interview with Paul Scheerer. Retrieved from https://fisheries.org/2015/07/qa-the-success-story-of-the-oregon-chub-an-interview-with-paul-scheerer/

Bastian, C. T., Keske, C., McLeod, D., & Hoag, D. (2017). Landowner and land trust agent preferences for conservation easements: Implications for sustainable land uses and landscapes. *Landscape and Urban Planning, 137*(1), 1–13.

Bean, M. J. (2017). Endangered species safe harbor agreements: An assessment working paper. Madison, WI: Sand County Foundation, Environmental Policy Innovation Center.

Bean, M. J., Bonnie, R., Male, T., & Searchinger, T. (2003). Private lands opportunity: The case for conservation incentives. Washington, DC: Environmental Defense.

Brown, G. M., & Shogren, J. F. (1998). The economics of the Endangered Species Act. *Journal of Economic Perspectives, 12*(3), 3–20.

Donlan, C. J. (Ed.). (2015). *Proactive strategies for protecting species: Pre-listing conservation and the Endangered Species Act.* Berkeley, CA: University of California Press.

Doremus, H. (2006). Lessons learned. In D. Gogle, J. Scott, & F. Davis (Eds.), *The Endangered Species Act at thirty* (Vol. 1, pp. 195–207). Washington, DC: Island Press.

Evans, D. M., Che-Castaldo, J. P., Crouse, D., Davis, F. W., Epanchin-Niell, R., Flather, C. H. . . . Williams, B. K. (2016, Winter). Species recovery in the United States: Increasing the effectiveness of the Endangered Species Act. *Issues in Ecology, 20*, 1–28.

Field, B. C., & Field, M. K. (2013). *Environmental economics: An introduction* (6th ed.). New York: McGraw-Hill, Irwin.

Fischer, P. (2004). Listening to landowners: Case studies from Oregon's Willamette Valley. West Linn, OR: Defenders of Wildlife and Willamette Basin Conservation Project.

Groves, C. R., Kutner, L. S., Stoms, D. M., Murry, M. P., Scott, J. M., Schafale, M., . . . Pressey, R. L. (2000). Owning up to our responsibilities: Who owns lands important for biodiversity? In B. Stein, L. Kutner, & J. Adams (Eds.), *Precious heritage: The status of biodiversity in the United States* (pp. 275–300). New York: Oxford University Press.

Lamothe, P., & Peterson, A. (2007). 2006 annual report. Candidate Conservation Agreement with Assurances for the fluvial Arctic grayling in the Upper Big Hole River. Retrieved from https://web.archive.org/web/20141123203430/http://dnrc.mt.gov/Default.asp

Langpap, C. (2004). Conservation incentives programs for endangered species: An analysis of landowner participation. *Land Economics, 80*(1), 375–388.

Langpap, C., Kerkvliet, J., & Shogren, J. (2018). The economics of the U.S. Endangered Species Act: A review of recent developments. *Review of Environmental Economics and Policy, 12*(1), 69–91.

Li, L. W., & Male, A. (2013). Protecting unlisted species: Assessing and improving candidate conservation agreement with assurances. White Paper Series. Washington, DC: Defenders of Wildlife.

Lueck, D., & Michael, J. (2003). Preemptive habitat destruction under the Endangered Species Act. *Journal of Law and Economics, 46*, 27–60.

Matsumoto, S., Pike, C., Turner, T., & Wah, R. (2003). The citizen's guide to the Endangered Species Act. Earth Justice. Retrieved from https://www.animallaw.info/article/detailed-discussion-great-apes-under-endangered-species-act

Montana Fish, Wildlife and Parks. (2006). Candidate conservation agreement with assurances for fluvial Arctic grayling in the Upper Big Hole River. Retrieved from https://www.fws.gov/mountain-prairie/es/species/fish/grayling/CCAA_June2006.pdf

Montana Fish, Wildlife and Parks. (2018). Candidate conservation agreement with assurances for the Arctic greying in the Centennial Valley, Montana. Retrieved from https://www.fws.gov/mountain-prairie/pressrel/2018/20180412_CVArcticGrayling_CCAA.pdf

Nielson, J. (2015). CCAA senior research essay. Department of Applied Economics. Corvallis, OR: Oregon State University.

O'Keefe, K. (1994). A "new American land ethic": Utilizing the Endangered Species Act to settle land disputes. *Florida State University Law Review, 21*(3), 1031–1059.

Paez, S. (2009). Preventing the extinction of candidate species: The lesser prairie-chicken in New Mexico. *Natural Resources Journal, 49*(2), 525–582.

Parker, D. (2004). Land trusts and the choice to conserve lands with full ownership or conservation easements. *Natural Resources Journal, 44*(2), 483–518.

Parker, D. (2008). The effects of public funding systems on the success of private conservation through land trusts. In T. Anderson, L. Huggins, & T. Power (Eds.), *Accounting for Mother Nature: Changing demands for her bounty* (pp. 167–194). Stanford, CA: Stanford University Press.

Ruhl, J. B. (2004). Endangered Species Act innovations in the post-Babbittonian era: Are there any? *Duke Environmental Law and Policy Forum, 14*(2), 419–440.

Salzman, J., & Thompson, B.H. (2017). *Environmental law and policy* (4th ed.). St. Paul, MN: Foundation Press.

Scott, D., Gogle, D., & Davis, F. (2006). Introduction. In D. Gogle, J. Scott, & F. Davis (Eds.), *The Endangered Species Act at thirty* (Vol. 1, pp. 3–15) Washington, DC: Island Press.

Shaffer, M., Watchman, L., Vickerman, S., Casey, F., Dewey, R., Snape, W., . . . Ferris, R. (2006). Proactive habitat conservation. In D. Gogle, J. Scott, & F. Davis (Eds.), *The Endangered Species Act at thirty* (Vol. 1, pp. 286–295). Washington, DC: Island Press.

Sorice, M. G., Oh, C., Gartner, T., Sniekus, M., Johnson, R., & Donlan, D. (2013). Increasing participation in incentive programs for biodiversity conservation. *Ecological Applications, 23*(5), 1146–1155.

Stumpf, J. M. (2008). The last best fish: Will conservation and consensus save Montana's Arctic grayling? (Unpublished graduate thesis). University of Montana, Missoula.

Tennant, L., Magee, J., Cayer, E., & McCullough, A. (2011). Conservation of fluvial Arctic grayling: A candidate conservation agreement in the Big Hole River, Montana. Conference of American Fisheries Society 140th Annual Meeting. Retrieved from www.researchgate.net/publication/267895575_Conservation_of_Fluvial_Arctic_Grayling_a_Candidate_Conservation_Agreement_in_the_Big_Hole_River_Montana

US Court of Appeals, Ninth Circuit. (2018). *Center for Biological Diversity, et al. vs. Zinke, et al.* No. 16-35866. DC. No. 2:15-cv-00004-SHE. Opinion. Retrieved from https://www.govinfo.gov/content/pkg/FR-2020-07-23/pdf/FR-2020-07-23.pdf

US Fish and Wildlife Service. (2003). Draft candidate conservation agreements with assurances handbook. Washington, DC: US Fish and Wildlife Service.

US Fish and Wildlife Service. (2005). Working together: Tools for helping imperiled wildlife on private lands. Washington, DC: US Fish and Wildlife Service.

US Fish and Wildlife Service. (2016, December). Endangered and threatened wildlife and plants; review of domestic and foreign species that are candidates for listing as endangered or threatened; annual notification of findings on resubmitted petitions; annual description of progress on listing actions. Washington, DC: US Fish and Wildlife Service.

US Fish and Wildlife Service. (2018). Endangered and threatened wildlife and plants; Enhancement of survival permit application; Centennial Valley Arctic grayling candidate conservation agreement with assurances and draft environmental assessment. Retrieved from www.govinfo.gov/content/pkg/FR-2018-05-29/pdf/2018-11367.pdf

US Fish and Wildlife Service. (2019, December). Endangered and threatened wildlife and plants; Review of domestic and foreign species that are candidates for listing as endangered or threatened; annual notification of findings on resubmitted petitions; annual description of progress on listing actions. Retrieved from www.federalregister.gov/documents/2019/10/10/2019-21478/endangered-and-threatened-wildlife-and-plants-review-of-domestic-and-foreign-species-that-arenotification

US Fish and Wildlife Service. (n.d.). Endangered Species Act of 1973 as amended through the 108th Congress. Washington, DC: US Fish and Wildlife Service.

US Fish and Wildlife Service & National Marine Fisheries Service. (1999a). Announcement of final policy for candidate conservation agreements with assurances. *Federal Register, 64*, 32726–32736.

US Fish and Wildlife Service & National Marine Fisheries Service. (1999b). Announcement of final policy for safe harbor agreements. *Federal Register, 64*, 32717–32725.

Wilson, R. (2014). Western states worry decision on bird's fate could cost billions in development. *Washington Post*. Retrieved from www.wasingtonpost.come/blogs/goybeatwp/2014/05/11western-states=worry-decision-on-birds-fate-could-cost_billions_in_development/

Wright, J. B., & Czerniak, R. J. (2000). The rising importance of voluntary methods of land use control in planning. *Journal of Planning Education and Research, 19*(4), 419–423.

Young, A. (2001, July 15). Federal marshals close Klamath Falls head gate. *Seattle Times*. Retrieved from https://archive.seattletimes.com/archive/?date=20010715&slug=klamath15m

16 Preserving endangered species

Communicating, reporting, and framing the extinction crisis

Eric Freedman, David B. Sachsman and Sara Shipley Hiles

Three dripping wet brown bear cubs – Kopala, Garincha, and Chofika – paced in the rain along the fence of their muddy temporary enclosure, waiting for wildlife ecologist and zoologist Khatia Basilashvili to feed them. She is a scientist working for the Tbilisi Zoo at a 120-acre site in a hilly area on the outskirts of the capital of the Republic of Georgia, a mountainous country in the South Caucasus. The cubs were to be the first participants in a pioneering program designed to rebuild the country's shrinking population of critically endangered brown bears (*Ursus arctos*). The plan, which drew little public attention when adopted, was to release them into Tbilisi National Park the following summer or early fall when berries and other forest foods would be plentiful.

According to Georgia's Red List, which is based on an evaluation of species status by the Commission of Endangered Species at the Academy of Sciences of Georgia, the 20th century witnessed the brown bear's disappearance from the country's floodplain forests and grasslands, limiting their current habitat to mountain forests and savannah-type landscapes. The Red List notes their attraction for eco-tourists and scientists, as well as their status as an "important character in Georgian national folklore."

Poachers pose the greatest threat to brown bears because they shoot mother bears, orphaning cubs, as happened to Kopala, Garincha, and Chofika. Most poachers aren't arrested, and those who do get busted usually end up with a warning or small fine. "Every spring they're killing mother bears and taking cubs. After one or two weeks, they realize the cubs are bears, not toys" and want to get rid of them because they're too much trouble or growing too big, Basilashvili explains.

Georgia's leading environmental investigative journalist, Tsira Gvasalia, has reported on impunity for poachers. In an article for the independent investigative news website ifact.ge (2017), she wrote:

> One evening in the southern Georgia village of Mlate, six hunters drove out into the forest. They parked and then scattered, looking for game. One of them, Akaki Chagunava, shot twice and killed a brown bear. . . . They got caught by local environment inspectors and were taken to court. The

> six hunters denied planning to kill a bear. "We could not imagine that Chagunava would kill a bear," said fellow hunter Gocha Kapianidze. "We were hunting for rabbits." The other hunters used their right to remain silent. The judge declared them and Kapianidze innocent and returned their guns.

Her investigation of judicial leniency, which also appeared in the English-language weekly newspaper *Georgia Today* and was cited by the online news site civil.ge, examined court records in 390 illegal hunting and fishing cases and found only five classified as criminal or killing a protected species. All five cases ended with plea bargains, and "no poacher paid a huge fine and nobody was sent to prison." The other cases were handled as minor administrative charges. "Poachers got back their guns and illegal fishing nets in 58 cases; in 203 cases, violators received only verbal warnings and no fines" (Gvasalia, 2017).

In addition to poaching, habitat fragmentation jeopardizes brown bears as new roads, development, and construction break up their territory. That makes it harder for males to travel from group to group to breed and for bears to hide and find food safely.

Before Basilashvili began her work, brown bear rescue in Georgia received little scientific attention. She read studies done elsewhere and identified experts to advise her in designing a rehabilitation program. When the cubs arrived at the zoo in the spring of 2018, the experts "decided we need to raise these bear cubs so they can survive naturally," not keep them in captivity forever, she says. They can live 20–30 years in the wild and twice as long in captivity. "They were stressed all day long. They were crying. It was creepy," Basilashvili tells a visitor. But when she opened the door to their enclosure and let them outside, "they were following me like I was their mother bear and they were my sons. They needed to play, sleep outside, climb trees, eat leaves or grass." If she hid behind a tree, "they cried for me."

At 4 months old, the cubs moved into a temporary space near separate areas for wolves, adult bears, and a shy Eurasian lynx. The zoo built a solid fence around their enclosure to help prepare them for eventual release. Food was dropped over the fence so they wouldn't see people and wouldn't associate food with people in their minds. To help them recognize what foods grow in the wild, Basilashvili collected apples, greens, and berries from the forest and hung them on trees in their enclosure. The cubs were to have GPS tracking collars around their necks so scientists would know exactly where they were. If the plan worked, more waves of translocated brown bears were expected to follow, making Kopala, Garincha, and Chofika forebears – so to speak – of a replenished population.

However, the release never happened for several reasons: lack of funds and relevant staff and "deficiency of readiness" by government agencies, and the cubs "are not capable of living in nature any more," according to Basilashvili (Basilashvili, 2021).

The proposed translocation and the country's poaching problem received some media attention. The zoo did not issue any press releases, but director Zurab Gurielidze often talked about the project on television programs and in interviews (Gurielidze, 2018). Overall, the Georgian press pays little attention to environmental news. A study (Freedman, 2020) of news media in Georgia attributed that sparsity of coverage to a series of factors, starting with the environment's lack of priority among media owners. Other factors were a shortage of journalists, reporters' lack of substantive knowledge about science and the environment, and their difficulty getting access to information in a country rated as only "partly free" by the democracy-building NGO Freedom House. That study also cited low public interest in the topic.

Communication about threatened and endangered species can bring science to the attention of policy makers and the citizenry, helping to set the agenda for discourse and decision-making (Dearing & Rogers, 1996). This is the goal of journalists like Tsira Gvasalia, but they do not always succeed. Poaching continues to be a problem in Georgia.

The following sections examine public reactions to the translocations and reintroductions of endangered species, communication about the role of zoos and aquariums in endangered species conservation, how news organizations report on and frame relevant events and policies, the effectiveness of celebrity activists, and how social media posts, documentaries, and videos can advance or impede science-based public awareness and action. Finally, the chapter discusses potential trends in communication about endangered species.

Translocations and reintroductions

Reintroductions and translocations are valuable management tools in the recovery of endangered and threatened species, especially to maintain genetic diversity, but many take place under the media radar with little or no news coverage. Nevertheless, which species to "save" and where to put them may be driven by a species' popularity instead of their threat status, according to Díaz, Anadón, Tella, Giménez, and Pérez (2018), who studied 174 Spanish conservation projects. Such projects, they observe, may be shaped by technical factors, such as feasibility and threat status, and by "the general awareness of the public and politicians on the necessity of these actions . . . Popularity, the degree of public knowledge or notoriety of a species, influences conservation decisions through its effects on the success of awareness and fund-raising campaigns" (p. 1420).

Conflict has long been a widely recognized news value among journalists and their media organizations (Parks, 2019). Thus, it is not unexpected for a controversial translocation proposal to draw headlines and evening news broadcast time, as shown by well-publicized debates over reintroducing wolves to areas where they had been extirpated or nearly extirpated. In some cases, those government and NGO activities anger farmers and ranchers who fear that returning apex predators will decimate their livestock. Press coverage, in turn, generates voter pressure on lawmakers and agency officials. In other locations, the debate is more philosophical, focusing on the fundamental question of the justification and morality of "genetic rescues," whether human activity caused or contributed to that species' downward spiral. In such situations, a key challenge for the news media is explaining to the public the intersection of science and societal values as determinants or influencers of government actions.

In the case of the Georgian bear cubs, not everyone in the conservation community was enamored with the proposed translocation, especially because it involved animals that may be dangerous to people. "Translocation is a very tricky thing," says Irakli Shavgulidze, the chair of the governing board of the Georgian eco-NGO NACRES (the Noah's Ark Center for the Recovery of Endangered Species), which was not involved in the translocation project. "The whole essence and purpose of any problem animal capture is it takes it somewhere else. . . . You don't want to displace the problem. If the problem goes with the animal, what's the point?" (Shavgulidze, 2018).

In an interview, Shavgulidze called Tbilisi National Park a poor choice because it's so close to the capital city, and he sounded skeptical about the cubs' ability to survive because they already had extensive human contact. "It's already too late as far as I'm concerned." And referring to potential problems with the park's proximity to Tbilisi and the cubs' survivability, Shavgulidze said, "The risks are so high they outweigh any rescue of individual animals and can damage the future for these species."

Not all objections to reintroduction or translocation are based on fear of predation or worries about human safety. Sometimes the underpinning science comes under question. For example, when the US Department of Agriculture proposed allowing genetically engineered (GE) American chestnuts to be planted, an advocacy group called the Campaign to STOP GE Trees objected and announced its opposition in a press release distributed to environmental reporters. The project was intended to replace a species that had been virtually wiped out by a fungal blight, killing an estimated 4 billion trees and leaving them what the American Chestnut Foundation calls technically or functionally extinct. The advocacy group that objected to the project cited a lack of long-term risk assessments, the involvement of major timber and chemical companies in the project, lack of consent from Indigenous landowners, and uncertainty that the new trees would prove blight-resistant. "Not only would this tree be the first ever GE forest tree approved for unrestricted planting in North America, it would also be the first genetically modified organism designed to spread into ecosystems," said the group, which has attracted press coverage over the years from a variety of news outlets, including the *Washington Post*, Politico, and public radio stations (Campaign to STOP GE Trees, 2020).

One significant translocation success story occurred in the United States when the endangered – and photogenic – black-footed ferret faced extinction from human-induced activities and fatal non-native diseases such as sylvatic plague. In 1986, only 18 ferrets were known to exist in the wild, all in

Wyoming. Scientists captured them "after conservationists failed to keep the last-known population alive in the wild" (Fritts, 2020) and set up a breeding program that now conducts annual releases. By 2018, the US Fish & Wildlife Service (USFWS) had released several hundred at 30 locations in eight Western US states, Canada, and Mexico in a program that has received widespread news media attention. Meanwhile, USFWS (n.d., p. 2) cautions that the black-footed ferret "remains one of the most endangered mammals in North America" due to disease, habitat loss, conversion of native grasslands to farming, and the eradication of prairie dogs, whose burrows are the ferrets' only housing and which constitute the largest part of the ferrets' diet. In large part because of media attention – including videos from *National Geographic* and PBS – and USFWS public education and public relations activities, black-footed ferrets are popular attractions at zoos, beneficiaries of continuing government-funded research and habitat protection, stars of a museum "ferret cam," and are the focus of a nonprofit ferret rescue organization.

Not all population restoration efforts are as successful. Consider the caribou, the subject of multiple translocations since 1924 in North America. In the "failed" efforts, as Leech, Jelinski, DeGroot, and Kuzyk (2017, p. 810) put it, predation, disease, human activity, and poaching were primarily to blame. One less-than-successful project translocated endangered mountain caribou into northern British Columbia, Canada. Those animals from outside the area failed to adopt "the predator avoidance strategies of resident caribou, rendering them more vulnerable to predation" (p. 809).

In a study of reintroduction efforts for 64 bird species and subspecies, Destro, De Marco, and Terribile identified 13 drivers that hurt reintroductions and reinforcement of their populations, most commonly predation, which they said has an "even more worrying" impact on islands than on the mainland (2018, p. 71). Others included "unexpected dispersal movement" – meaning the birds don't stay where they are supposed to stay – disease, and human impacts such as hunting, poisoning, and collisions with motor vehicles and power lines.

Translocations and release initiatives often generate conflicts, such as the conflict between those who support the release of wolves into the wild and ranchers worried about their livestock, and between those who support the conservation of African elephants in reserves and nearby villagers worried about damage to their crops.

Such conflicts can be highly political, as occurred in the US when the Trump administration dropped plans favored by conservation groups to reintroduce grizzlies in the North Cascade Mountains. The administration's decision followed vociferous opposition to the plans from farmers, property owners, and homeowners (Geranios, 2020).

Coogan and colleagues wrote of a "dichotomy in societal responses to scientific news stories" about Canadian and US grizzly bears (2019, p. 93). On one hand, they see strong positive backing for news stories about topics that are "popular" among the public, such as research about conservation challenges for the bears, settlement and recreation patterns, road density, and rail lines. At the same time, Coogan and colleagues point to public skepticism "and even outright denial as to the quality and value of scientifically acquired knowledge relating to 'unpopular' conservation issues" (p. 93), such as delisting the bears under the Endangered Species Act and allowing trophy hunting in British Columbia. Coogan and colleagues proffer several possible explanations for the dichotomy between positive and negative societal reactions regarding conservation research:

- People may tend to back research supporting their own agendas or opinions about the bears.
- Opinions are often passionate about grizzlies and their conservation.
- "[T]he actual science behind the news story (and not the simplified and often misinterpreted media message) is complex."
- Agendas and opinions are amplified and the science possibly distorted via proliferation through social media.

Beyond translocation situations, it is not unusual for the interests of landowners to actually or seemingly clash with conservation strategies and mandates intended to protect endangered species. In some situations, compromises get implemented. In other instances, governmental regulations appear necessary to resolve conflicts. The question, then, for environmental communicators, scientists, policy makers, and wildlife managers, is how to weigh competing values and vested interests. This is how American philosopher, essayist, and poet Wendell Berry addressed conflicts between coyotes and sheep in the Western United States, between those who use nature directly and those who use it indirectly:

> The fact is, people need both coyotes and sheep, need a world in which both kinds of life are possible. Outside the heat of conflict, conservationists probably know that a sheep is one of the best devices for making coarse foliage humanly edible and that wool is ecologically better than synthetic fibers, just as most shepherds will be aware that wild nature is of value to them and not lacking in interest and pleasure.
>
> The usefulness of coyotes is, of course, much harder to define than the usefulness of sheep. Coyote fur is not a likely substitute for wool, and, except as a last resort, most people don't want to eat coyotes. The difficulty lies in the difference between what is directly useful. Coyotes are useful indirectly, as part of the health of nature, from which we and our sheep alike must live and take our health. The fact, moreover, may be that sheep and coyotes need each other, at least in the sense that neither would prosper in a place unfit for the other.
>
> (Berry, 1987, pp. 10–11)

There are additional complications to consider as well because such conflicts may harm other endangered species. For example, measures taken by cattle ranchers in the Upper Galilee region of Israel to protect their herds from endangered gray wolves also killed griffon vultures through poisoning and mountain gazelles through habitat fragmentation by fences (Becker & Farja, 2017).

The role of zoos and aquariums

Aquariums and zoos – which generally have their own public relations, communications, and public education and outreach staff – have long collaborated with partners such as conservation groups and public agencies to help species recover through breeding, relocation, research, population monitoring, and public outreach and education activities (Che-Castaldo, Grow, & Faust, 2018; da Silva et al., 2019). Nevertheless, government agencies have underappreciated the conservation role of zoos, says Scott Carter, the chief life sciences officer at the Detroit (United States) Zoo (Carter, 2020).

The underappreciation of zoos and aquariums may be fueled in part by the way the media frame their conservation and education functions, one study suggests. Maynard's (2018) qualitative content analysis of North American magazine articles about zoos and aquariums found that animal welfare was the most common frame, while "other frames, such as conservation and education, were less prevalent . . . than expected, since zoos and aquaria portray themselves to be leaders in these fields" (p. 177). "Negative" coverage includes stories questioning the "moral and scientific basis" of zoos, the "morality of captivity," lack of "empathy or respect" for animals, "inappropriate reinforcements of human dominance over all other species," and "problematic animal welfare, husbandry, and politics" (p. 177).

Carter sees a "real push" to get all accredited zoos to ramp up their conservation efforts and commit more broadly "beyond just breeding and releasing," such as site monitoring, field research, and cooperative management of species with other institutions. "There are so few species from whom breeding and release is a possibility," he said in an interview.

The Detroit Zoo's conservation focus is on non-charismatic species, primarily amphibians. "Two decades ago, we identified a crisis coming for amphibian species. They were not charismatic and not getting a lot of attention," Carter says. On average, amphibians are the most threatened animal group, according to the International Union for Conservation of Nature. At the same time, Carter says their releases are easier to carry out than other types of fauna, there are protected areas set aside for them, they produce many offspring, and they're easily transported. The zoo's National Amphibian Conservation Center breeds and releases endangered Wyoming toads and rare striped newts, for example. It also maintains "assurance populations" of other species such as the Panamanian golden frog and dusky gopher frog, both critically endangered, which are being kept in captivity as insurance against extinction until there are suitable locations to release and protect them in the wild. "There is no habitat available [for them] for a variety of reasons. Not everything has an immediate release date," Carter says.

The Detroit Zoo's online "press room" issues news releases on its breeding activities, not only for amphibians such as the critically endangered Puerto Rican crested toad, but also for other species, such as its program for the endangered piping plover, a North American shorebird. A press release about the hatching and release of 39 plover chicks (Detroit Zoological Society, 2020) generated stories in major and regional newspapers across Michigan, at least one commercial radio station, and a newspaper about 900 miles from the zoo in Savannah, Georgia, where the piper winters over. It got further publicity through Facebook and on the websites of a variety of conservation, ornithology, and environmental organizations. Jonathan Slaght (2020), in his account of five years of field work designing a conservation plan for the endangered fish owl in Russia's Far East, writes:

> Fish owls lack the name recognition and star power of Amur tigers. Conservation organizations devote entire fundraising drives to Amur tigers, with millions of dollars a year raised for their protection. With fish owls, funding is limited to whatever grants [my collaborator] and I have time to cobble together.

His book drew wide media attention: *The Times of London* named Slaght's vivid account one of the best nature books of 2020, it was nominated for a National Book Prize, and it received reviews in the *Wall Street Journal*, the *Guardian*, the *New York Times*, the *Budapest Times*, the *Spectator*, and other media outlets.

Aside from questions of science and survival, difficulties may arise when a species' recovery relies on captive breeding and reintroduction, according to Melstrom, Salau, and Shanafelt (2019). Discussing the endangered black-footed ferret that depends on captive breeding for survival, the researchers point to tradeoffs between the cost and duration of the breeding program and how long the population spends in recovery and in the wild. Yet, that is not the type of pragmatic question the popular press is apt to cover.

The practicality and ethics of selective breeding and cloning may be controversial. The cloning process uses genetic material to produce individual organisms with identical DNA. In the case of extinct species, DNA sometimes can be extracted from bones and preserved tissue. In South Africa, the director of a project to use DNA and selective breeding of the plains zebra to bring back its extinct relative, the quagga, acknowledges, "There are a lot of detractors who are saying you can't possibly put back the same as what was here" (Page & Hancock, 2016). Referring to proposals to clone the Ice Age wooly mammoth, which was wiped out about 4,000 years ago, the Detroit Zoo's Carter (2020) says, "What can you do with them? Bringing back an extinct species – de-extinction – and keeping it in captivity seems a bit questionable."

And when Australian researchers voiced the possibility of cloning the extinct thylacine, or Tasmanian tiger, they acknowledged that the odds of success were tiny – "8% at best," an Australian Museum evolutionary biologist on the project said. "It's a very long shot." Even if the science were to work, a key question was where the resurrected carnivorous marsupial would live, now that much of its traditional habitat in more heavily populated Tasmania had been developed or destroyed. Could it survive in the wilds? Or would it be doomed to permanent captivity? (Freedman, 2002, p. 18) Asked about the morality of people "playing God," the museum biologist, who said he's a religious man, shook his head and replied, "I don't know what that means. It's never been explained to me. This is not anything I've prayed about." And the museum director said, "My response is that people played God when we exterminated the animal in the first place."

News coverage and framing of relevant events and policies

In an age of rising concerns about climate change, shrinking fauna and flora habitat, and worsening extreme weather events, the news media are paying more attention to threats of species extinction, although one can question whether that attention is enough to arouse public demands for remedial action or enough to prod policy makers to take meaningful steps to respond to the crisis. Yet news organizations' financial and personnel resources to increase coverage are limited. That's because the novel and unusual attract attention, and novelty is a basic element of newsworthiness in science coverage (Molek-Kozakowska, 2017). As Tbilisi Zoo scientist Basilashvili observed, "When something happens" – as when an elephant underwent surgery at the zoo in 2018 – "it's a big story [in] every TV and newspaper." That story drew coverage by the British Broadcasting Corporation and Reuters, as well as by Georgian media (Freedman, 2020).

The press has widely reported some back-from-the-brink good news:

- The birth of the first cloned colt of a Przewalski's horse, a species that has survived mainly in zoos for the past four decades, may make it possible to successfully breed and infuse genetic diversity into the species' small remaining population (Johnson, 2020a).
- Scientists in Djibouti rediscovered a species of elephant shrew, last scientifically recorded in the 1970s (Briggs, 2020).
- Using a remotely operated vehicle for an underwater expedition, US government scientists discovered a new species of carnivorous comb jelly near Puerto Rico, 2½ miles below sea level (Johnson, 2020b).
- The USFWS has been able to downlist or delist recovering or recovered species under the Endangered Species Act. Among them: the Okaloosa darter and West Indian manatee, which the agency moved from endangered to threatened, and the Louisiana black bear and bald eagle, which were delisted (US Geological Survey, 2018).
- The endangered Tasmanian devil has returned to the Australian mainland for the first time in three millennia, beneficiaries of a reintroduction program (Bittel, 2020). Meanwhile, the transmission rate of the fatal

Tasmanian devil facial tumor disease, a cancer that is responsible for an 80% decline in its population, has slowed and the animal's decimated population may be recovering (Patton et al., 2020).

On a broader scale, California Academy of Sciences (2019) researchers identified 71 animal and plant species for the first time in 2019. Among them were sea slugs, eels, corals, and lizards in locations as diverse as Croatian caves, the waters off the Falkland Islands, a Brazilian national park, the Chihuahuan Desert in Mexico, and a national marine sanctuary off the coast of San Francisco.

And there's much disappointing news reported as well:

- WWF's 2020 Living Planet Index says humans are destroying biodiversity "at a rate unprecedented in history" (p. 6), with the biggest declining populations of mammals, birds, reptiles, amphibians, and fish in Latin America and the Caribbean. The study also found the decline in biodiversity is moving faster among freshwater species than those in oceans and forests.
- The critically endangered Sumatran rhino was extirpated from Malaysia in 2019, leaving tiny remnant populations in Indonesia (Katz, 2019).

It is not surprising to see "bad news" stories fall into several predictable patterns: this species on the brink; that species on the brink; this species miraculously pulled back from the brink; that species miraculously pulled back from the brink. News organizations are under constant pressure to build and retain audiences, and it is questionable how well such coverage will hold the attention of readers, viewers, and listeners.

Framing is one way journalists exercise news judgment in deciding which elements of a controversy or event to include or omit, and which words and images to use. As Entman (1993, p. 52) explained:

> Framing essentially involves selection and salience. To frame is to select some aspects of a perceived reality and make them more salient in a communicating text, in such a way as to promote a particular problem definition, causal interpretation, moral evaluation, and/or treatment recommendation for the item described.

Gan, Teo, and Detenber observed, "The fact that framing might lead audiences to have different reactions is an important implication for political communication" (2005, p. 443). In the context of the Endangered Species Act, communications researchers examined how the three largest US television networks covered the conflict between competing "procut" and "prosave" advocates regarding the old-growth forest home of the threatened Northern spotted owl in the Pacific Northwest:

> [B]oth sides endeavored to frame the issue in ways that they perceived would foster public sympathy. For the procut side, this meant defining it as

> a narrow conflict between an obscure bird and a way of life that supports thousands of people. Seen in those terms, the people losing their jobs seem more sympathetic than the owl, no matter how cute the latter may be. For the prosave side, the goal was to widen the terms of debate, framing it as an unfortunate collision between the need to preserve the ancient and magnificent public resources of the old-growth forest and an industry that was shrinking for a wide range of reasons.
>
> (Liebler & Bendix, 1996, p. 61)

Many stories frame environmental disputes as "bad" versus "good" when it comes to nearby Indigenous lands and sacred places, including petroleum pipeline projects in the United States, construction of a wall to deter unauthorized migration along the United States–Mexico border, and fracking projects in Canada. In those situations, news reports talk about the projects' environmental and cultural threats. However, one shortcoming of much news coverage of endangered species is the lack of attention to the question of what constitutes one group's "good" versus another's "good," such as situations that pit conservationists against local populations, such as native peoples. Cheng offered this example: North of the Arctic Circle in Norway where the golden eagle was once rare, the raptor's return threatens the reindeer that many Indigenous Sami depend on for their livelihoods. Eagle numbers plummeted due to over-hunting and poisoning, loss of habitat, and other human disturbances, but have rebounded due to protections under international treaty and Norwegian law. "For reindeer, the raptors are a particular threat during calving season. . . . It is not uncommon for golden eagles to kill far more reindeer than they can consume in one sitting" (Cheng, 2020, p. 30).

There also have been valid critiques of how poorly traditional and mainstream media cover Indigenous and local peoples' perspectives on endangered species. Those critiques often sharply attack journalists for the limited, often elite, types of news sources interviewed for such stories. To illustrate, Julia Cantzler (2007) analyzed 268 articles and letters to the editor in US and international newspapers about a well-publicized conflict between the Makah Tribe in Washington state and anti-whaling activists. The tribe wanted to resume its 2,000-year tradition of hunting gray whales for ceremonial and traditional purposes, which may be in conflict with international agreements. She found that some published accounts included rhetoric from anti-hunting sources "suggesting the savage or uncivilized nature of past or present tribal practices" with "an implication that whaling is . . . immoral" and that "draw upon negative stereotypes of American Indians to bolster support for their cause." Other statements by anti-hunting crusaders "demonstrate an obvious lack of sensitivity to tribal customs and cultural traditions," call into question the tribe's integrity, and fail "to recognize the importance of the [t]ribe being able to determine its own cultural destiny" (pp. 504–505).

Celebrities in the spotlight

Scholars define "celebrity endorsement" as "active participation by a celebrity in delivering campaign messages" and an "environmental conservation campaign" as "the organized delivery of messages that focus on actions benefiting the natural environment and/or biodiversity, including both protection against a specific threat and conservation generally," according to Olmedo et al. (2020, p. 3). To illustrate, finning – the process of cutting off shark fins and tossing the sharks back into the sea – is salient and controversial, not only because it jeopardizes shark populations but also because of traditional cultural mores and high financial stakes. Nor is the problem merely theoretical. In 2020, for example, fishing company Hamada Suisan Co., Ltd. pleaded guilty and agreed to a $126,000 fine and forfeited another $119,000 because the Indonesian crew of its fishing boat had harvested fins from protected sharks, including whitetip, silky sharks, and bigeye thresher sharks. The company told a federal court in Hawaii that it was implementing a corporate compliance plan with new training and stronger cooperation with regulators to prevent finning (Kelleher, 2020).

For environmental communications, one key question is what value the involvement of celebrities – dubbed "celanthropists" – brings to efforts to protect endangered species. Turner (2016, p. 811) wrote that "environmental politics has been 'spectacularized' through the media . . . in the deployment of celebrity." Other examples include performers Alicia Silverstone and Ellen DeGeneres on behalf of veganism. Turner notes:

> Even though we can name numerous occasions when the celanthropist has worked in precisely the way the textbooks might tell us – raising the profile of an issue and determining how it is regarded by the public – the enlistment of the celebrity as a means of generating attention for environmental issues by no means guarantees success.
>
> (2016, p. 812)

Meanwhile, a meta-analysis by Olmedo et al. (2020) of celebrity-endorsed environmental campaigns in nine countries concluded that it is unclear whether such campaigns contribute to conservation efforts. They said the lack of measurable objectives, theory of change, critical evaluation, and outcome indicators made it impossible to determine whether the campaigns' outcomes can be attributed to the celebrity endorsement.

Beyond the "mainstream" press: social media posts, documentaries, and videos

As researchers have found, science-based documentaries and videos can draw attention to extinction and other environmental challenges, such as climate change. As Morgan Richards wrote, "Wildlife documentary has come to assume a key role in the public understanding of science and environmental issues, generating popular awareness and helping to shape public engagement with environmental politics and conflict" (2013, p. 173). Noting that videos can increase awareness of rare species and create sensitivity towards the environment, Kleinhenz and Parker (2017) produced three educational films about the Endangered Species Act, each focused on one animal or plant species. After screening them for 10th-graders with "behavioral challenges," they concluded that "each video demonstrated significant effects on student content retention and one video caused a positive attitude shift in students towards the Endangered Species Act" (p. 46), making the case for using nature documentaries to teach students about endangered species.

Increasingly now, social media posts supplement, and perhaps in some cases supplant, more traditional nature documentaries as environmental communication channels. The ubiquitous social media content, whether rooted in science or in misinformation, ungrounded opinion, and speculation, highlights that information on endangered species is not the sole domain of mainstream media, scholarly journals, traditional documentaries, and specialized print, online, and broadcast environmental news outlets.

Social media systems are a two-edged sword, and this trend has serious implications for threatened and endangered species. Some of those implications may be positive, such as raising awareness of the need for conservation action, generating revenue for local communities, and helping conservationists mobilize to oppose development projects, resource extraction operations, and other activities that jeopardize habitat, access to food, and protection from human and non-human predators.

However, there are negative implications too, as when social media posts entice large numbers of people to visit ecologically critical habitats and interact with wild animals. Large-scale visitation is among the factors that can impair successful breeding and survival. The growth of wildlife tourism to see the charismatic, endangered, yellow-eyed penguins on the South Island of New Zealand contributes substantially to the local economy, but the human presence at their breeding grounds "can cause delays in beach landings, reducing the amount of food available to chicks, leading to reduced chick growth and fledgling mass, and, eventually, decreased survival rates" (Bell, Young, Seddon, & van Heezik, 2020, p. 469). Thus, the risk is great when one of the world's rarest penguins, with an estimated population of fewer than 1,700 pairs, draws 40,000–50,000 tourists annually to a single one of its breeding areas.

A study of 282 YouTube videos about mountain gorilla tourism in Africa reflects that dilemma (Otsuka & Yamakoshi, 2020). Gorillas and people appeared in close proximity or physical contact with each other in 40% of them, in violation of international regulations. Not surprisingly, those videos attracted more "likes" and views than the others. As Otsuka and Yamakoshi (2020, p. 3) put it:

> The ecotourism of endangered mountain gorillas has attracted many tourists and functioned as a key component of their conservation. There might

> be expectations on the part of tourists to observe or interact . . . in close proximity and such expectations may have been engendered by the contents of social media in this Information Age. However, the risk of disease transmission . . . is a large concern and it is important to maintain a certain distance . . . to minimize risk.

Also on the negative side, online videos and websites may facilitate trade in endangered species. An analysis of 847 YouTube videos showing 173 pet species of amphibians found footage of two species that were listed on the Convention on International Trade in Endangered Species (CITES) Appendix I. Overall, more than 15% of the species were listed on CITES Appendix II, although the researchers could not determine how many had been captured in the wild and how many had been bred in captivity (Measey et al., 2019). A study of a Hong Kong–based pet trade website reported how "the Internet is being exploited as a medium for illegal trade" of CITES-listed chelonian turtles (Wong, Lee, Cheung, Lam, & Tang, 2020). Even more troubling were the findings of a study of social media showing exotic pets in the Middle East: Eighty-five percent of the species were listed in CITES Appendices I and II, including great apes, birds of prey, and big cats (Spee, Hazel, Grande, Boardman, & Chaber, 2019).

Some experts contend that "the multi-billion dollar wildlife trade is the primary reason for species extinction" (Hruby, 2020, p. A9). Here is an example from Spee et al. (2019): Popularization of exotic pets on social media "has led to questionable impacts on . . . demand and threats to species conservation" (p. 1). The annual global value of illegal trafficking of exotic species is an estimated $20 billion, according to the study. The social media accounts in that study were on Instagram, Facebook, Twitter, and YouTube and belonged to "highly influential persons from oil-rich Middle Eastern regions," namely, members of royal families.

Bias is an accusation long leveled against the news media – whether mainstream independent media, government-controlled or government-owned media, or alternative media. Environmental communication about endangered species is not exempt from such criticism. One study critiqued "simplistic and inaccurate coverage of trophy hunting in [print and online] mainstream newspapers" (Hart et al., 2020, p. 1333) and the conflation of legal trophy hunting with poaching. It went on to say:

> The dominant media narrative is beguilingly simple: "well-loved animals are on the brink of extinction, people pay to kill them, this is horrible, and if we stop it we will save species". . . . The clear message in most coverage is that stopping trophy hunting would have immediate positive effects on conservation. However, the reality is not so simple.
>
> (p. 1333)

Trophy hunting can provide economic incentives to protect species and habitats, the study said, and it criticized "false 'extinction narratives'" when the sport helps conserve the critically endangered black rhino and the near-threatened white rhino, markhor goats, and argali sheep. In the United States, a decline in hunters has meant less revenue for a federal fund that supports conservation projects on public lands – many of which directly benefit endangered species – as well as reduced state revenue from hunting licenses and fees (Sellers, 2020).

Looking into the crystal ball

The editors of this book cannot look into a crystal ball, read tea leaves, or spread out tarot cards to predict the future of environmental communication about endangered species and its impact on public policy. However, they can speculate about key factors and trends that will influence that future.

First, despite the best efforts of governments, industries, and individuals – efforts that have not yet and may never come to fruition – anthropogenic climate change will continue its disruptive effects on endangered fauna and flora. Consider this example that highlights the intersection of climate-related habitat disruption: A disease associated with anomalously warm water along the Pacific Coast between Alaska and Mexico is being blamed for a 90.5% decline in sunflower sea stars – the death of as many as 5.75 billion of them since 2013. The implications go beyond the impact on the starfish themselves: Their near disappearance allowed the sea urchin population to explode, and that has triggered a decline in many marine kelp forest environments, researchers say (Nair, 2020). Global warming is also likely to continue and perhaps exacerbate the frequency and severity of extreme weather, further damaging ecosystems.

Second, growing, relocating, and changing populations will increase pressure on wildlife and habitats. Some of that pressure will be climate-induced as people move inland from coastal areas inundated by rising sea levels. Some of the pressure on wildlife and habitats will be development of roads, hydroelectric projects, and other infrastructure that further fractures the land. Some of the pressure on wildlife and habitats will be caused by unexpected events, such as the COVID-19 global pandemic with its crushing, tragic economic as well as public health consequences. Writing about the coronavirus pandemic, Nuwer (2020, p. 31) said:

> Many terrestrial protected areas are . . . experiencing an uptick in bushmeat poaching, much of it likely driven by people out of work. In May 2020, in Sri Lanka, a rare black leopard – one of only three recorded in the past decade – was killed by a snare trap probably intended for deer. In India, researchers found that the number of animals illegally killed for meat nearly doubled during the COVID-19 national lockdown, compared with the same time last year.

Third, science-based decision-making on conservation and other environmental policies and laws will continue to face a barrage of misinformation, disinformation, "fake news," and unjustified attacks on researchers, at least in some countries. As Lees, Attwood, Barlow, and Phalan (2020, p. 1440) commented, "Denial of scientific evidence and rejection of scientific methods are not new phenomena, but represent an increasingly serious problem, especially when driven by politically well-connected and well-funded antagonists seeking to sabotage evidence-based policy for political and/or financial gain." They identified three categories of extinction denialism:

- literal denial that overlooks and underestimates recent extinctions;
- interpretive denial by framing the extinction crisis as a historical problem and only a minor contemporary challenge;
- implicatory denial, suggesting that technology and targeted interventions will overcome extinction.

As evidence of the long life of science denials, we see anti-vaxxers wage a seemingly endless war against vaccines that have proven safe and effective, foregoing inoculations that would protect their families, neighbors, and communities from serious preventable diseases. The still-extant flat earth movement, evolution non-believers, and anti-fluoride activists provide additional evidence.

Next, we anticipate escalating changes in communication technologies and public expectations of the media, whether traditionally mainstream, specialized, or "social." Promises that "everyone" can become a publisher – or at least a communicator to the public – are increasingly real, yet too often accompanied by media illiteracy and scientific illiteracy. Blogs and social media posts – or their successors – can become tools for raising consciousness among policy makers, policy shapers, and the citizenry around the world. At the same time, they can be proliferators of "fake science," exploitation, and misinformation that impedes rather than advances protection of endangered species.

Finally, we expect that coverage of endangered species will continue going forward. Reflecting the media's traditional roles in framing controversies and in setting public agendas, a key question is which types of media attention will drive more public interest that may, in turn, drive action by governments, industries, and other decision-makers and policy implementers. Responsible, science-based news and information from journalists and other communicators could be part of the solutions.

References

Basilashvili, K. (2018, November 16). Interview with E. Freedman. Tbilisi, Republic of Georgia.

Basilashvili, K. (2021, January 5). Email to E. Freedman.

Becker, N., & Farja, Y. (2017). The cattle-wolf dilemma: Interactions among three protected species. *Environmental Management*, *59*, 175–188.

Bell, G., Young, M. J., Seddon, P. J., & van Heezik, Y. (2020). Effects of unregulated visitor access on chick fledgling mass and survival in yellow-eyed penguins. *Wildlife Research*, *47*(6), 468–475.

Berry, W. (1987). *Home economics*. San Francisco, CA: North Point Press.

Bittel, J. (2020, October 5). Tasmanian devils return to mainland Australia for first time in 3,000 years. *National Geographic*. Retrieved from www.nationalgeographic.com/animals/2020/10/tasmanian-devils-return-to-mainland-australia/

Briggs, H. (2020, August 18). Elephant shrew rediscovered in Africa after 50 years. BBC. Retrieved from www.bbc.com/news/science-environment-53820395

California Academy of Sciences. (2019, December 5). Academy scientists describe 71 species in 2019 (Press Release). Retrieved from www.calacademy.org/press/releases/academy-scientists-describe-71-species-in-2019

Campaign to STOP GE Trees. (2020, August 18). Researchers' plan to release genetically engineered American chestnut trees in forests will set dangerous precedents if approved (Press Release). Retrieved from https://stopgetrees.org/usda-release/

Cantzler, J. M. (2007). Environmental justice and social power rhetoric in the moral battle over whaling. *Sociological Inquiry*, 77(3), 483–512.

Carter, S. (2020, October 20). Telephone interview with E. Freedman.

Che-Castaldo, J. P., Grow, S. A., & Faust, L. J. (2018). Evaluating the contribution of North American zoos and aquariums to endangered species recovery. *Science Reports, 8*. doi:10.1038/s41598-018-27806-2

Cheng, K. C. (2020). In the Arctic, a conflict over hoof and claw. *Earth Island Journal, 35*(30), 24–32.

Coogan, S. C. P., Coops, N. C., Janz, D. M., Cattet, M. R. L., Kearney, S. P., Stenhouse, G. B., & Nielsen, S. E. (2019). Towards grizzly bear population recovery in a modern landscape. *Journal of Applied Ecology, 26*, 93–99.

da Silva, R., Pearce-Kelly, P., Zimmerman, B., Knott, M., Foden, W., & Conde, D. A. (2019). Assessing the conservation potential of fish and corals in aquariums globally. *Journal for Nature Conservation, 48*, 1–11.

Dearing, J. W., & Rogers, E. M. (1996). *Agenda setting*. Thousand Oaks, CA: Sage.

Destro, G. F. G., De Marco, P., & Terribile, L. C. (2018). Threats for bird population restoration: A systematic review. *Perspectives in Ecology and Conservation, 16*, 68–73.

Detroit Zoological Society. (2020, August 27). Piping plover captive rearing program celebrates most successful year to date (Press Release). Retrieved from https://detroitzoo.org/press-release/piping-plover-captive-rearing-program-celebrates-most-successful-year-to-date/

Díaz, M., Anadón, J. D., Tella, J. L., Giménez, A., & Pérez, I. (2018). Independent contributions of threat and popularity to conservation translocations. *Biodiversity Conservation, 27*, 1419–1429.

Entman, R. M. (1993). Framing: Toward clarification of a fractured paradigm. *Journal of Communication, 43*(4), 51–58.

Freedman, E. (2002, Winter). Cut from history: An abandoned Tasmanian zoo tells the haunting tale of an endling. *EJ Magazine*, pp. 16–18.

Freedman, E. (2020). Not a priority: Barriers to environmental reporting in the Republic of Georgia. *Applied Environmental Education and Communication*. doi:10.1080/1533015X.2020.1749732.

Fritts, R. (2020, September 7). An invisible enemy: The battle to save black-footed ferrets from the plague. *The Guardian*. Retrieved from www.theguardian.com/environment/2020/sep/07/an-invisible-enemy-the-battle-to-save-black-footed-ferrets-from-the-plague-aoe

Gan, F., Teo, J. L., & Detenber, B. H. (2005). Framing the battle for the White House: A comparison of two national newspapers' coverage of the 2000 United States presidential election. *International Communication Gazette, 67*(5), 441–467.

Geranios, N. K. (2020, July 7). Feds scrap plans to reintroduce grizzlies to North Cascades. Associated Press News. Retrieved from https://apnews.com/article/4454db59750f8b0b3bb21895ead3d662

Gurielidze, Z. (2018, December 5). Interview with E. Freedman. Tbilisi, Republic of Georgia.

Gvasalia, T. (2017). Georgian judges go soft on poachers. *IFact*. Retrieved from https://ifact.ge/en/poachers/

Hart, A. G., Cooney, R., Dickman, A., Hare, D., Jonga, C., Johnson, P. K., . . . Somerville, K. (2020). Threats posed to conservation by media misinformation. *Conservation Biology*. doi:10.1111/cobi.13605

Hruby, D. (2020, November 28). A life spent caring for (and poaching) the world's rarest birds. *New York Times*.

Johnson, L. M. (2020a, September 12). A clone of the endangered Przewalski's horse is born of DNA saved for 40 years. CNN. Retrieved from https://lite.cnn.com/en/article/h_addf9b35fcb0760d709c20ba2f431729

Johnson, L. M. (2020b, November 28). NOAA scientists discover a new species of gelatinous animal in the waters near Puerto Rico. CNN. Retrieved from www.cnn.com/2020/11/28/us/new-species-of-comb-jelly-trnd/index.html.

Katz, B. (2019, November 26). Sumatran rhinos are now extinct in Malaysia. *Smithsonian*. Retrieved from www.smithsonianmag.com/smart-news/sumatran-rhino-now-extinct-malaysia-180973648

Kelleher, J. C. (2020, October 9). Fishing boat owner pleads guilty in shark finning case. Associated Press News. Retrieved from https://apnews.com/article/sharks-fish-archive-honolulu-hawaii-12b262d67e4c878aa180f9a5e72cb6f2

Kleinhenz, P. N., & Parker, M. S. (2017). Videos as a tool to increase understanding and support for the Endangered Species Act. *Applied Environmental Education & Communication, 16*(1), 41–55.

Leech, H., Jelinski, D. E., DeGroot, L., & Kuzyk, G. (2017). The temporal niche and seasonal differences in predation risk to translocated and resident woodland caribou (*Rangifer tarandus caribou*). *Canadian Journal of Zoology, 95*, 809–820.

Lees, A. C., Attwood, S., Barlow, J., & Phalan, B. (2020). Biodiversity scientists must fight the creeping use of extinction denials. *Nature Ecology & Evolution, 4*, 1440–1443. doi:10.1038/s41559-020-01285-z

Liebler, C. M., & Bendix, J. (1996). Old-growth forests on network news: News sources and the framing of an environmental controversy. *Journalism & Mass Communication Quarterly, 73*(1), 53–65.

Maynard, L. (2018). Media framing of zoos and aquaria: From conservation to animal rights. *Environmental Communication, 12*(2), 177–190.

Measey, J., Basson, A., Rebelo, A. D., Nunes, A. L., Vimercati, G., Louw, M., & Mohanty, N. P. (2019). Why have a pet amphibian? Insights from YouTube. *Frontiers in Ecology and Evolution*, 7(52).

Melstrom, R. T., Salau, K. R., & Shanafelt, D. W. (2019). The optimal timing of reintroducing captive populations into the wild. *Ecological Economics, 156*, 174–184.

Molek-Kozakowska, K. (2017). Communicating environmental science beyond academia: Stylistic patterns of newsworthiness in popular science journalism. *Discourse & Communication, 11*(1), 69–88.

Nair, R. (2020, December 16). 5.7b sunflower sea stars have died in past half-decade, bringing species to brink of extinction. Canadian Broadcasting Corporation. Retrieved from www.cbc.ca/news/canada/british-columbia/sunflower-sea-star-decline-1.5844674

Nuwer, R. (2020). Nature is returning. *Sierra Magazine*. Retrieved from www.sierraclub.org/sierra/2020-6-november-december/feature/nature-returning-during-covid-coronavirus-pandemic

Olmedo, A., Milner-Gulland, E. J., Challender, D. W. S., Cugnière, L., Dao, H. T. T., Nguyen, L. B., . . . Veríssimo, D. (2020). A scoping review of celebrity endorsement in environmental campaigns and evidence for its effectiveness. *Conservation Science and Practice, 2*(10). doi:10.1111/csp2.261

Otsuka, R., & Yamakoshi, G. (2020). Analyzing the popularity of YouTube videos that violate mountain gorilla tourism regulations. *PLoS One, 15*(5). doi:10.1371/journal.pone.0232085

Page, T., & Hancock, C. (2016, January 27). Zebra cousin went extinct 100 years ago. Now it's back. CNN. Retrieved from www.cnn.com/2016/01/25/africa/quagga-project-zebra-conservation-extinct-south-africa/

Parks, P. (2019). Textbook news values: Stable concepts, changing choices. *Journalism & Mass Communication Quarterly, 96*(3), 784–810.

Patton, A. H., Lawrance, M. F., Margres, M. J., Kozakiewicz, C. P., Hamede, R., Ruiz-Aravena, M., . . . Storfer, A. (2020). A transmissible cancer shifts from emergence to endemism in Tasmanian devils. *Science, 370*, 6522. doi:10.1126/science.abb9772

Richards, M. (2013). Greening wildlife documentary. In L. Libby & B. Hutchins (Eds.), *Environmental conflict and the media* (pp. 171–185). New York: Peter Lang.

Sellers, F. S. (2020, February 2). Hunting is "slowly dying off," and that has created a crisis for the nation's many endangered species. *Washington Post*. Retrieved from www.washingtonpost.com/national/hunting-is-slowly-dying-off-and-that-has-created-a-crisis-for-the-nations-public-lands/2020/02/02/554f51ac-331b-11ea-a053-dc6d944ba776_story.html

Shavgulidze, I. (2018, November 19). Interview with E. Freedman. Tbilisi, Republic of Georgia.

Slaght, J. C. (2020). *Owls of the Eastern ice: A quest to find and save the world's largest owl*. New York: Farrar, Straus and Giroux.

Spee, S. B., Hazel, S. J., Del Grande, E., Boardman, W. S. J., & Chaber, A. L. (2019). Endangered exotic pets on social media in the Middle East: Presence and impact. *Animals, 9*(480).

Turner, G. (2016). Celebrities and the environment: The limits to their power. *Environmental Communication, 10*(6), 811–814.

US Fish & Wildlife Service. (n.d.). Black-footed ferret: Information sheet. Washington, DC: US Fish & Wildlife Service.

US Geological Survey. (2018). Collaborative species conservation. Retrieved from www.usgs.gov/news/collaborative-species-conservation

Wong, R. W. Y., Lee, C. Y., Cheung, H., Lam, J. Y. K., & Tang, C. (2020). A case study of the online trade of cites-listed chelonians in Hong Kong. *Chelonian Conservation and Biology, 19*(1), 95–100.

WWF. (2020). Living planet report 2020: Bending the curve of biodiversity loss. Retrieved from https://livingplanet.panda.org/en-us/

Index

Note: Numbers in **bold** indicate a table. Numbers in *italics* indicate a figure.